T0228213

Atomic Diffusion in
III−V Semiconductors

Atomic Diffusion in III–V Semiconductors

Brian Tuck

Department of Electrical and Electronic Engineering,
University of Nottingham

Adam Hilger, Bristol and Philadelphia

©IOP Publishing Ltd 1988

All rights reserved. No part of this publication may be reproduced, stored in a retrieval system or transmitted in any form or by any means, electronic, mechanical, photocopying, recording or otherwise, without the prior permission of the publisher.

British Library Cataloguing in Publication Data

Tuck, Brian, *1940–*
    Atomic diffusion in III–V semiconductors.
    1. Semiconductors. Diffusion
    I. Title
    537.6'22

    ISBN 0-85274-351-3

Library of Congress Cataloging-in-Publication Data

Tuck, Brian.
    Atomic diffusion in III–V semiconductors.

    Bibliography: p.
    Includes index.
    1. Semiconductors—Diffusion. I. Title.
    II. Title: Atomic diffusion in 3–5 semiconductors.
    III. Title: Atomic diffusion in three–five
    semiconductors.
    QC611.6.D5T79   1988   621.3815'2   88-4490
    ISBN 0-85274-351-3

Published under the Adam Hilger imprint by IOP Publishing Ltd
Techno House, Redcliffe Way, Bristol BS1 6NX, England
242 Cherry Street, Philadelphia, PA 19106, USA

Typeset by Mathematical Composition Setters Ltd, Salisbury
Printed in Great Britain by J W Arrowsmith Ltd, Bristol

Contents

Preface   vii

1   Gallium Arsenide and Friends   1

2   Elements of Diffusion   9
   2.1   The diffusion equations   9
   2.2   Analytical solutions to the diffusion equation   11
   2.3   Finite difference methods of solution   20
   2.4   Experimental techniques   24
   2.5   Analysis of results   27
   2.6   Interaction of defects   32
   2.7   The built-in field effect   38
   2.8   The external system   41
   2.9   Assessment of published data   45

3   Diffusion of Shallow Donors, including Group IV   47
   3.1   Sulphur   47
   3.2   Selenium and tellurium   61
   3.3   Tin   62
   3.4   Silicon and germanium   69

4   Shallow Acceptors, especially Zinc   75
   4.1   The substitutional–interstitial mechanism   75
   4.2   Zinc in GaAs   78
   4.3   Zinc in GaP   91
   4.4   Zinc in InP   100
   4.5   Zinc in other compounds   105
   4.6   Cadmium in InP   108
   4.7   Cadmium in other compounds   112
   4.8   Diffusion of other group II elements   114

vi   *Contents*

5   Diffusion of Transition Elements   117
    5.1   Chromium in GaAs   118
    5.2   Manganese in GaAs   142
    5.3   Iron in GaAs   153
    5.4   Cobalt in GaAs   155
    5.5   Iron and chromium in InP   156

6   Other Fast Diffusers   161
    6.1   Silver in InP   161
    6.2   Silver in GaAs   177
    6.3   Silver in InAs and GaP   185
    6.4   Diffusion of gold   187
    6.5   Diffusion of copper   189

7   Self-diffusion and Related Phenomena   193
    7.1   Diffusion in GaAs   194
    7.2   InP and InAs   201
    7.3   Self-diffusion in InSb   203
    7.4   Diffusion in GaSb   204
    7.5   Diffusion in superlattices   210

References   223

Index   233

Preface

Modern semiconductor devices rely on the technologist's ability to introduce predetermined amounts of dopant into precisely defined regions of a chip. It follows that any movement of impurities within the semiconductor must be carefully controlled or, at the very least, predictable. As devices become smaller, this becomes more difficult to achieve and more accuracy is required in the modelling of the diffusion processes. When dealing with diffusion phenomena in semiconductors, it is tempting to feed in the appropriate boundary conditions and to assume that one or other of the well known solutions of Fick's law will describe the movement of the atoms. This is usually wrong, and much of this book is devoted to explaining why. It is not that there is anything wrong with Fick's law, merely that, in general, more than one process is occurring at the same time and they invariably interact.

Following initial chapters on III–V semiconductors and basic diffusion theory, the book provides a critical review describing our present state of knowledge of diffusion in the III–Vs. The literature in this subject is enormous, and some degree of selection has been inevitable. I have tried to concentrate on those studies, both experimental and theoretical, which have furthered our understanding of the atomic mechanisms involved. The subject of self-diffusion, which has been somewhat neglected by experimentalists, is dealt with at some length, with attention being paid to the fascinating results which have been obtained recently on MBE samples.

A stern effort has been made to reduce the amount of jargon used to an absolute minimum, largely on principle but also because I am acutely aware of the fact that potential readers will have been reared in a wide variety of scientific disciplines. In this connection, perhaps I should confess that nomenclature has proved a serious problem throughout the writing of the book. The attentive reader will notice that some symbols mean different things in different places. S, for instance is used to mean 'substitutional atom', but also 'sulphur', similarly, P can stand for 'pressure' or 'phosphorus'. A completely unambiguous notation could have been devised, but it would have involved a truly horrific system of subscripts and superscripts. I decided it was better to compromise in this matter, but have tried to be meticulous about saying what the symbols mean within a given section.

Brian Tuck
Nottingham, 1988

Chapter 1

Gallium Arsenide and Friends

The III–V semiconductors are compounds made between elements of group III, such as gallium, indium, aluminium, and group V atoms such as arsenic, phosphorus, antimony. All which are of interest for device manufacture have the zincblende crystal structure, in which each group III atom has four group V atoms as nearest neighbours, and *vice versa*, as shown in figure 1.1. Both silicon and germanium crystallise in essentially the same form; when it occurs in an element it is called 'diamond' structure. The distance between adjacent atoms varies by about 20 per cent for different members of the group, from 0.236 nm for GaP to 0.280 nm for InSb. The band-gaps vary by a much greater factor, from 0.2 eV (InSb) to 2.5 eV (AlP), and this proves to be one of the most useful properties of the III–Vs. Information on the band-gaps, together with the melting-points of the materials, is given in table 1.1.

Interest in the materials dates from the 1950s when, because of its high electron mobility, GaAs was first seriously considered as a possible alternative to germanium and silicon for transistor applications. About the same time it was reported that a GaAs p–n junction gives out light when biassed in the forward direction and the first light emitting diodes were made. In the 1960s an entirely new line of enquiry was initiated when Gunn discovered that microwave oscillations can be obtained from bulk GaAs when it is subjected to a high electric field. All of these aspects have proved fruitful in the intervening years and the interest has spread from GaAs to the whole family of semiconductors. More recently, major advances have taken place in technology which permit new devices to be made, and interest in the III–Vs is greater than ever before.

As with silicon, the first requirement is to produce high quality

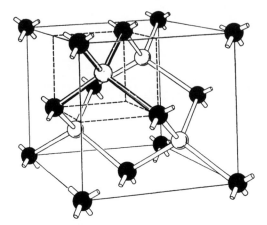

Figure 1.1   Zincblende structure.

Table 1.1

Material	Melting point (K)	Energy gap (eV)
GaAs	1510	1.44
GaP	1740	2.26
GaSb	985	0.72
InAs	1215	0.35
InP	1340	1.35
InSb	798	0.18
AlAs	2010	2.16
AlP	2800	2.45
AlSb	1333	1.50

single crystals with impurity levels which are determined by the crystal grower rather than by chance. Most of the III–Vs can be produced as large bulk crystals, but only GaAs and InP are currently made in commercial quantities. The two major methods of growth are the horizontal Bridgeman and the Czochralski techniques, with the latter now the more important, producing single crystals of good circular cross-section in kilogramme amounts. Slices from these bulk crystals are usually used as substrates on which further layers are grown. Crystals can be produced within a wide range of electrical conductivity; in particular, very highly resistive material can be made. Both GaAs and InP, for example, are commonly grown with resistivity in the region of 10^7 ohm cm. The availability of this 'semi-insulating' material for use as substrates represents a major advantage for the III–V group. No equivalent can be produced in silicon and this leads to complications in silicon technology when electrical isolation is required. Problems have arisen in the use of semi-insulating substrates, however, some of which are due to diffusion. They will be considered in a later chapter on the diffusion of transition elements.

Three types of technique are used to grow layers on substrates. They may be summarised as follows:

(1) Chemical vapour deposition (CVD). In this method, vapours are passed over the substrate inside a vessel of some kind. The temperatures within the vessel are arranged so as to cause a reaction to take place on the substrate surface and the semiconducting compound is deposited.

(2) Liquid phase epitaxy (LPE). The substrate is brought into contact with a supersaturated solution. After equilibration, the temperature is slowly reduced, causing the compound to come out of solution and grow onto the surface.

(3) Molecular beam epitaxy (MBE). This technique is the simplest in principle, but probably the most complicated in practice since it takes place in ultra-high vacuum. Beams of group III and group V atoms are directed at a heated substrate. The beams are generated by separate source cells which contain extremely pure materials. The flux of atoms is controlled by the temperature of the source cell.

Because all of the III–Vs of interest have the same crystal structure, the possibility arises of depositing one material on another. The difficulty here, of course, is that in general different compounds will have different values of lattice constant. In the III–V group, however, there are several examples of compatibility; GaAs and AlAs, for example, have almost the same lattice constant. Other compatible pairs are GaSb/AlSb and GaP/AlP. An extra degree of freedom is therefore gained in the production of semiconductor junctions and a number of important devices are now made using heterojunctions rather than the more conventional homojunctions.

Further flexibility exists due to the fact that solid solutions can be made which are mixtures of two or more of the basic III–Vs. The solution $GaAs_xP_{1-x}$ exists over the complete range $0 < x < 1$, and can be prepared reproducibly for any required value of x. This means that the technologist can make to order a semiconductor with a band-gap of any value between those of GaAs and GaP; indeed, since he has available to him the whole range of III–V semiconductors, he can prepare one with any value between 0.2 eV and 2.5 eV. This proves to be a powerful facility in the production of a number of modern devices.

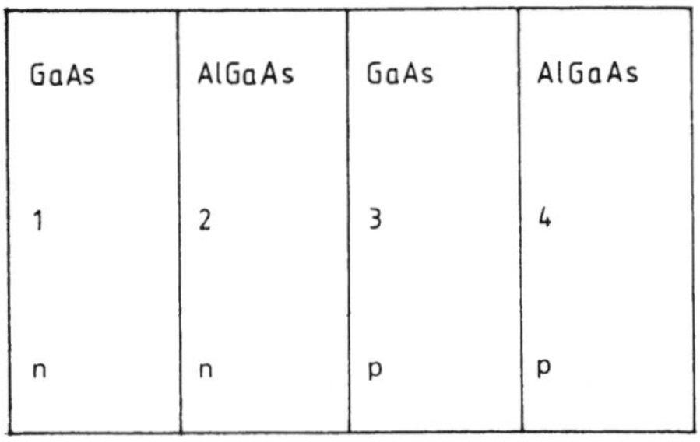

GaAs
AlGaAs
GaAs
AlGaAs
1
2
3
4
n
n
p
p

Figure 1.2 Structure of double-heterojunction laser. Layer 3 is the active layer.

1
2
3
4
$h\nu$
GaAs
AlGaAs
GaAs
AlGaAs

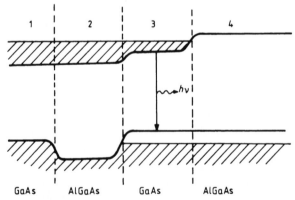

Figure 1.3 Band diagram for laser of figure 1.2 with applied forward bias.

A good example of the use both of solid solutions and of heterojunctions is provided by the double heterojunction laser, shown diagramatically in figure 1.2. The device consists essentially of layers of $Al_xGa_{1-x}As$ and GaAs, grown on a GaAs substrate. The band-gap of the solid solution is greater than that of GaAs for all $x > 0$; a typical value of x for the device might be 0.3–0.4. The energy band diagram for the structure is shown in figure 1.3. The active region of the laser, i.e. the layer in which the electrons and holes recombine to produce photons, is the p-GaAs region labelled 3 in the figures. This layer is made very thin, of the order of 1 μm or less, and the effect of the two heterojunctions is to concentrate the electrons and holes within it, ensuring the population inversion which is essential to laser action. Confinement of carriers is not of itself sufficient to achieve optimum laser efficiency, however. It is also necessary to confine the light so that as little as possible leaks into the adjacent layers. Herein lies the real subtlety of the double heterojunction laser; as well as having different values of band-gap, the two materials also have different refractive indices. An efficient wave-guide structure therefore exists, restricting the light within the p-GaAs layer. The energy of the photons emitted from the device corresponds to the band-gap of GaAs, 1.4 eV, i.e. the light output is in the near infra-red with wavelength about 900 nm.

The most important use of solid state lasers at present is in the communications industry. Transparent optical fibres are used to transmit light signals over considerable distances. It is obviously highly desirable in such a system for the wavelength of the source to correspond to the minimum in the attenuation curve for the fibre optic material. While the laser described above transmits at a fairly suitable wavelength for fibres currently in use, it does not have the optimum value, which is in the region of 1.3 μm. A semiconductor layer of band-gap appropriate to this wavelength could easily be made using a solid solution of GaAs and InAs. Unfortunately, the lattice constant of this material would not correspond to that of either of the easily available substrate materials (GaAs and InP). The problem has been overcome by using quaternary solid solutions rather than ternary. The addition of the fourth element to the compound permits an extra degree of freedom so that, within limits, the lattice constant of the material can be chosen as well as the band-gap. The quaternary chosen is $Ga_xIn_{1-x}As_yP_{1-y}$. The idea is expressed in figure 1.4, in which the energy gap of the available materials is plotted against lattice spacing. The shaded area between the four binary compounds InAs, InP, GaAs, GaP shows the permitted range. It can be seen that it is possible to choose values of *x* and *y* to give a compound which has the required band-gap and can be grown on an InP substrate. The resulting device is shown in figure 1.5.

Recent advances in the production of ultra-thin epitaxial layers have opened up a whole new range of possibilities. Experts in MBE growth, for instance, tell us that is is now possible to grow films

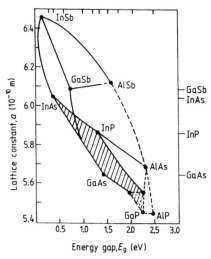

Figure 1.4 Variation of energy gap with lattice constant in the III–V compounds.

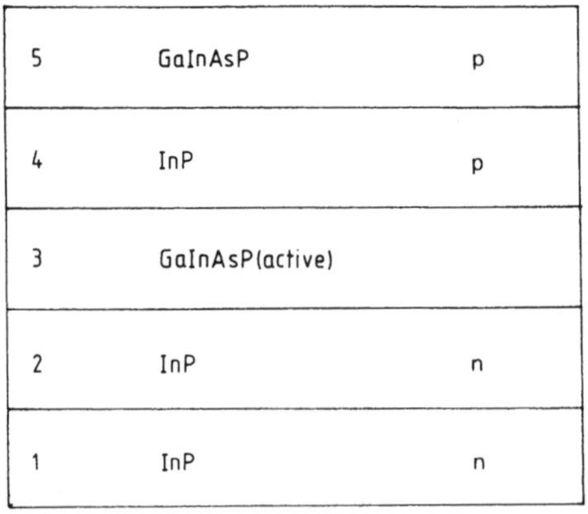

5	GaInAsP	p
4	InP	p
3	GaInAsP(active)	
2	InP	n
1	InP	n

Figure 1.5 Structure of GaInAsP double-heterojunction laser. Layer 1 is the substrate; light is produced in layer 3.

of one or two monolayers. Conventional band theory was not designed for this sort of thing, and a new area of low-dimensional physics is in the process of being developed. Already, remarkable experimental findings have been published. It is found, for instance, that if a structure is grown consisting of a number of thin layers which are alternately high and low band-gap material, interesting energy-level effects occur. If such a structure, usually called a 'multiple quantum well' (MQW) is grown using GaAlAs and GaAs as the two constituents, for example, it is found that optical transitions originating from the GaAs 'wells' have photon energy greater than the band-gap of GaAs. Thus, if the GaAs active layer in the laser of figure 1.2 is replaced by a MQW structure, the emission is of lower wavelength than before. The precise value of this wavelength depends on the thickness of the layers in the MQW; the thinner the layers, the lower the wavelength. If the thickness of the GaAs layers is reduced to 1.3 nm, the emission is at about 700 nm, which is in the visible region of the spectrum.

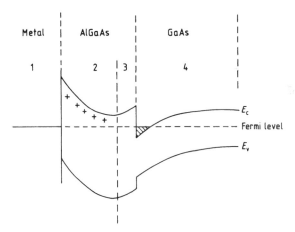

Figure 1.6 Energy band diagram showing two-dimensional electron gas (shaded in layer 4). Layer 2 is doped n-type, layers 3 and 4 are undoped.

Another important low-dimension effect is the so-called 'two-dimensional electron gas' which can be created at a heterojunction interface. An example can be seen in figure 1.6. The interface is made between a large band-gap semiconductor, GaAlAs, which has been doped n-type, and an undoped lower band-gap material, GaAs. The band structure is such as to create an energy well just inside the GaAs layer, which the electrons from the GaAlAs occupy. The electrons are therefore removed from the donors which gave rise to them. Electrons are able to move along the well (i.e. perpendicular to the plane of the figure) with very high mobility, since they are not subjected to ionised impurity scattering. The phenomenon is the basis of the potentially important 'high electron mobility' transistor, currently under development in many laboratories.

The degree of interest in the III–V semiconductors is now greater than ever before and it seems likely that the rate of progress of the last 30 years will be maintained into the forseeable future. In this short chapter an attempt has been made to give the reader the flavour of what is an exciting and fast-moving field of work. We will now limit ourselves to a consideration of one part of the story, namely the study of atomic diffusion in this group of materials.

Chapter 2

Elements of Diffusion

The subtleties of diffusion in semiconductors have been dealt with at some length in an earlier book (Tuck 1974). In this chapter, a rather more brief description is offered, giving only that theory which is required to make the following chapters comprehensible. In addition, an account is given of some of the more important experimental techniques, together with methods of analysis of results.

2.1 The Diffusion Equations

Diffusion occurs in the presence of a concentration gradient of mobile atoms. In general, the atoms will move in such a way as to remove the gradient. If the atomic jumps are random and independent of each other, it is fairly easy to show that the flux of atoms at any point is down the concentration gradient and proportional to it, i.e.

$$J = - D \nabla C \tag{2.1}$$

where J is the flux of atoms at the point x, y, z (a vector), C is the concentration at the same point and D is the constant of proportionality, usually called the diffusion coefficient. Equation (2.1) is known as Fick's law and is one of a class of laws (including Ohm's law, for instance) which says that effect is proportional to cause. It should be noted at the outset, however, that in the case of semiconductor diffusion, the progress of the atoms through the crystal is often not simple, and many examples of 'non-Fickian' diffusion occur.

The diffusion coefficient is invariably a function of temperature and the relationship

$$D = D_0 \exp\left(-\frac{Q}{kT}\right) \qquad (2.2)$$

often applies. The energy Q can be related to the free energy required for an atom to jump from one stable position in the crystal to the next.

Fick's second law is derived by applying a continuity condition to equation (2.1). Consider the flow of diffusing material through an element of volume $\Delta x \Delta y \Delta z$, as shown in figure 2.1. There are components of matter flow in all three directions; the y and z flows are omitted from the diagram for the sake of simplicity. The continuity argument states that the rate at which material is accumulating in the element must be equal to the rate at which it is flowing in, minus the rate at which it is leaving.

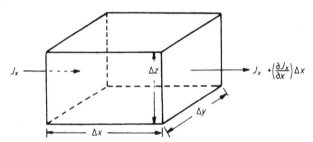

Figure 2.1   Flow of diffusing material through a volume element.

The rate at which it accumulates due to flow in the x-direction is

$$\Delta y \Delta z \left[J_x - J_x + \left(\frac{\partial J_x}{\partial x}\right)\Delta x \right] = -\frac{\partial J_x}{\partial x}\,\Delta x \Delta y \Delta z.$$

The rate at which it accumulates due to flow in all three directions is therefore

$$-\Delta x \Delta y \Delta z \left(\frac{\partial J_x}{\partial x} + \frac{\partial J_y}{\partial y} + \frac{\partial J_z}{\partial z}\right) = \frac{\partial C}{\partial t}\,\Delta x \Delta y \Delta z$$

where the right-hand side follows from the fact that $\Delta x \Delta y \Delta z$ is the volume of the element. Hence

$$\frac{\partial C}{\partial t} + \frac{\partial J_x}{\partial x} + \frac{\partial J_y}{\partial y} + \frac{\partial J_z}{\partial z} = 0. \qquad (2.3)$$

In an isotropic medium, we can substitute from equation (2.1) and get

$$\frac{\partial C}{\partial t} = \frac{\partial}{\partial x}\left(D\frac{\partial C}{\partial x}\right) + \frac{\partial}{\partial y}\left(D\frac{\partial C}{\partial y}\right) + \frac{\partial}{\partial z}\left(D\frac{\partial C}{\partial z}\right).$$ (2.4)

If D is a constant, this becomes

$$\frac{\partial C}{\partial t} = D\nabla^2 C.$$ (2.5)

In many practical cases it is necessary to consider diffusion in only one dimension. If there is a gradient only along the x-axis, equations (2.4) and (2.5) reduce to

$$\frac{\partial C}{\partial t} = \frac{\partial}{\partial x}\left(D\frac{\partial C}{\partial x}\right)$$ (2.6)

and

$$\frac{\partial C}{\partial t} = D\frac{\partial^2 C}{\partial x^2}.$$ (2.7)

Equation (2.7) is usually called Fick's second law, although it is often necessary to use equation (2.6) in the case of semiconductor diffusion. The relationship can also, of course, be expressed in terms of cylindrical or spherical co-ordinates; for semiconductor work, however, the Cartesian form is normally the most appropriate.

2.2 Analytical Solutions to the Diffusion Equation

The mathematical solutions to Fick's second law depend on the boundary conditions, which are determined by the physical conditions of the experiment in question. In this section problems are considered for which the diffusion coefficient is a constant; cases for which it is variable are dealt with later. Only a selection of the more important conditions which arise in semiconductor diffusion can be considered here. More comprehensive accounts of solutions of the diffusion equation for almost any conceivable situation are given by Crank (1956) and Carslaw and Jaeger (1959). In order to keep life simple we will consider only one-dimensional problems.

2.2.1   Thin film solution

It is a common experimental technique to use as diffusion source a very thin layer of diffusant deposited on the face of the semiconductor slice. The problem is to solve Fick's law subject to the condition that the total amount of diffusant in the system does not change with time of diffusion. It is simpler initially to consider the slightly different case of an infinite bar along the x-direction with a thin diffusant film situated at $x = 0$, perpendicular to the length of the bar. This problem is preferable because it is symmetrical about the origin. The bar is raised to a high temperature and diffusion takes place. It is a simple matter to show that after a time t, the distribution must be Gaussian:

$$C = \frac{A}{t^{1/2}} \exp\left(\frac{-x^2}{4Dt}\right). \qquad (2.8)$$

The expression obeys equation (2.7) and the boundary conditions of the experiment i.e. it is symmetrical with respect to $x = 0$, goes to zero as x goes to $\pm$ infinity for $t > 0$, and for $t = 0$ vanishes everywhere, except for $x = 0$, where it is infinite (it has been assumed that the thickness of the original film is zero). It remains to demonstrate, however, that the total amount of diffusing material does not change with time. Let this quantity be M; according to equation (2.8),

$$M = \int_{-\infty}^{\infty} \frac{A}{t^{1/2}} \exp\left(\frac{-x^2}{4Dt}\right) dx. \qquad (2.9)$$

Changing the variable to $v^2 = \dfrac{x^2}{4Dt}$, equation (2.9) becomes

$$M = \int_{-\infty}^{\infty} 2AD^{1/2} \exp(-v^2) dv \qquad (2.10)$$

and, using the well-known relation

$$\int_{-\infty}^{\infty} \exp(-v^2) dv = \pi^{1/2}$$

we have

$$M = 2A(\pi D)^{1/2}. \qquad (2.11)$$

Thus the amount of solute in the bar is constant with respect to time and the first condition is obeyed. The complete thin-film solution is therefore:

$$C = \frac{M}{2(\pi Dt)^{1/2}} \exp\left(\frac{-x^2}{4Dt}\right).$$ (2.12)

Figure 2.2 shows equation (2.12) for various times.

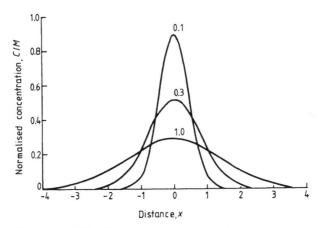

Figure 2.2   Thin-film solutions to the diffusion equation for different diffusion times, plotted as C/M against x. The numbers on the curves refer to different values of the quantity Dt.

Let us now return to the more realistic case in semiconductor diffusion of a thin film deposited on a 'semi-infinite' slice. Since the 'infinite' problem is symmetrical, we can slice it down the middle and take the solution for positive x. Physically this is reasonable, since figure 2.2 shows that at $x = 0$, dC/dx is zero at all times. It follows from Fick's first law that there is no flow of atoms across the origin, so the two halves of the problem are effectively insulated from each other. We must allow for the fact that in the infinite problem half the mass goes in one direction and half in the other. Thus if we split the infinite problem into two semi-infinite ones, the amount of material deposited on each of the surfaces is $\frac{1}{2}M$. The solution can be written:

$$C = \frac{\alpha}{(\pi Dt)^{1/2}} \exp\left(\frac{-x^2}{4Dt}\right)$$ (2.13)

where α is the amount of diffusant deposited on the surface of a semi-infinite specimen.

It is now possible to be a little more precise about what is meant in practice by a 'semi-infinite' specimen. Suppose it has a length l in the diffusion direction. At $x = l$, the diffusant may evaporate from the end of the specimen or may be reflected back into it, depending on the conditions of the experiment. In either case, the profile will be disturbed from that described by equation (2.13). Provided the concentration of diffusant at the far end is very small, this effect will be negligible, and the specimen can reasonably be called semi-infinite. Define this condition by stipulating that the concentration at $x = l$ must be no more than 1 percent of the value at $x = 0$, i.e.

$$\exp\left(\frac{-l_{\min}^2}{4Dt}\right) = 0.01$$

where $l_{\min}$ is the minimum length for a semi-infinite specimen. A value of $4.3(Dt)^{1/2}$ is obtained for $l_{\min}$, underlining the importance of the length $(Dt)^{1/2}$ which is the natural unit of length in this subject. In practice, diffusion into semiconductor slices is often carried out so as to make the penetration much less than the thickness of a slice, rendering the semi-infinite approximation valid. In similar fashion, a 'thin' layer can be defined as one of thickness less than or comparable to $(Dt)^{1/2}$.

2.2.2   Thick film solution

This is the case for which the thickness of a deposited layer of diffusant is not negligible compared to $(Dt)^{1/2}$. Again it is simpler to consider the case of the film sandwiched between two semi-infinite bars. Let the film occupy the space $-h < x < h$ and the bars occupy $-\infty < x < -h$, and $h < x < \infty$ respectively, so that the problem is symmetrical about $x = 0$. The boundary conditions may therefore be expressed

$$\left.\begin{array}{l} C = C' \quad -h < x < h \\ C = 0 \quad \text{elsewhere} \end{array}\right\} \quad \text{at } t = 0.$$

The simplest way to solve the problem is to imagine the thick film to be made up of many thin films. We can then sum the contributions from the films, using equation (2.12).

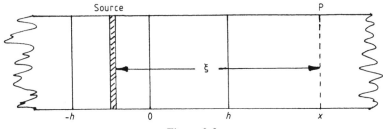

Figure 2.3

Figure 2.3 shows the bar and one of the thin-film sources. Consider the effect which this source has on some general plane P. Let P be situated at the point x, distance ξ from the source. The thickness of the source is $d\xi$ and it contains an amount of solute $C'd\xi$. From equation (2.12), the concentration of solute at P after diffusion time t is:

$$\frac{C'd\xi}{2(\pi Dt)^{1/2}} \exp\left(\frac{-\xi^2}{4Dt}\right). \tag{2.14}$$

It follows that the concentration at P due to all of the elemental thin film sources must be given by the sum from $\xi = x - h$ to $\xi = x + h$

$$C = \int_{(x-h)}^{(x+h)} \frac{C'}{2(\pi Dt)^{1/2}} \exp\left(\frac{-\xi^2}{4Dt}\right) d\xi. \tag{2.15}$$

Changing the variable to $\eta = \xi/2(Dt)^{1/2}$, this becomes

$$C = \frac{C'}{\pi^{1/2}} \int_{b'}^{b} \exp(-\eta^2)\, d\eta \tag{2.16}$$

where $b = (x + h)/2(Dt)^{1/2}$ and $b' = (x - h)/2(Dt)^{1/2}$.

This can be re-written

$$C = \frac{C'}{\pi^{1/2}}\left(\int_{0}^{b} \exp(-\eta^2)\, d\eta - \int_{0}^{b'} \exp(-\eta^2)\, d\eta\right). \tag{2.17}$$

We now have the distribution in the form of standard functions. The error function is defined as

$$\operatorname{erf} z = \frac{2}{\pi^{1/2}} \int_{0}^{z} \exp(-\eta^2)\, d\eta. \tag{2.18}$$

Clearly this function has the properties erf $z = -$ erf$(-z)$ and erf $0 = 0$. We also define the error function complement function erfc $z = 1 -$ erf z.

Substituting equation (2.18) into equation (2.17),

$$C = \tfrac{1}{2} C'(\text{erf } b - \text{erf } b')$$

or, using the above properties of the error function,

$$C = \tfrac{1}{2} C'(\text{erf } b + \text{erf } - b').$$

Substituting for b and b' gives the final result

$$C = \frac{C'}{2}\left(\text{erf } \frac{h+x}{2(Dt)^{1/2}} + \text{erf } \frac{h-x}{\partial(Dt)^{1/2}}\right). \qquad (2.19)$$

Both the erf and erfc functions have been extensively calculated and values are given in standard tables.

2.2.3   Pair of semi-infinite specimens

Consider a pair of semi-infinite bars, one occupying the space $-\infty < x < 0$, the other occupying $0 < x < \infty$. They are joined at the plane $x = 0$, which is perpendicular to the length of the bars. The first bar is homogeneously doped with solute to a concentration C', the other contains no solute. Apart from this, the two bars are identical. The temperature is raised to diffusion temperature at time $t = 0$. The boundary conditions are:

$$\left. \begin{array}{l} C = C' \text{ for } x < 0 \\ C = 0 \quad \text{for } x > 0 \end{array} \right\} \quad \text{at } t = 0.$$

The problem can be solved in much the same way as the thick film one in section 2.2.2. The doped bar can be considered as being made up of an infinite number of elemental thin films, from $x = -\infty$ to $x = 0$. Again we take some general point x and consider the concentration due to an elemental source at distance ξ (see figure 2.3). Equation (2.14) applies, and the amount at x due to all such films is found this time by integrating the expression from $\xi = x$ to $\xi = \infty$, i.e.

$$C = \int_x^\infty \frac{C'}{2(\pi Dt)^{1/2}} \exp\left(\frac{-\xi^2}{4Dt}\right) d\xi. \qquad (2.20)$$

Following the same line of reasoning as before, this becomes

$$C = \frac{C'}{2} \frac{2}{\pi^{1/2}} \int_b^\infty \exp(-\eta^2) \, d\eta \qquad (2.21)$$

where $\qquad b = x/2(Dt)^{1/2}.$

Thus

$$C = \frac{C'}{2} \left(\frac{2}{\pi^{1/2}} \int_0^\infty \exp(-\eta^2) \, d\eta - \frac{2}{\pi'} \int_0^b \exp(-\eta^2) \, d\eta \right) \qquad (2.22)$$

and, using the properties of the error function given in the previous section, we have

$$C = \tfrac{1}{2} C' (1 - \text{erf } b) = \tfrac{1}{2} C' \text{ erfc} \frac{x}{2(Dt)^{1/2}}. \qquad (2.23)$$

The error function complement curve is plotted in figure 2.4 in normalised form, using $2(Dt)^{1/2}$ as the unit of length.

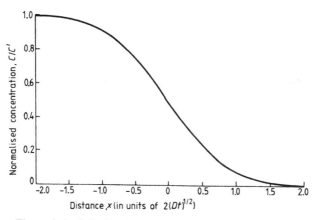

Figure 2.4 Solution for pair of semi-infinite specimens.

2.2.4 Constant surface concentration

The set of boundary conditions defined in the previous section for the two semi-infinite specimens would be difficult to realise in practice for a semiconductor system. One might imagine, for instance, growing a thick doped crystal onto an undoped thick substrate made of the same material, and then raising the couple to an elevated temperature for diffusion to occur. In general, however, the crystal growth process itself would be expected to take place at elevated temperature, so some diffusion would happen during growth, disturbing the required step condition. A very common experimental situation does occur, however, for which the boundary conditions of the previous section are relevant, namely that of a diffusion process in which the surface concentration is kept constant with time.

One of the features of the erfc curve is that the value of the function at $x = 0$ is unity for all time. Reference to equation (2.23) shows that at $x = 0$ the concentration is always $\frac{1}{2} C'$. It follows that for positive values of x, the problem described in section 2.2.3 is identical to one in which the concentration is maintained at $\frac{1}{2} C'$ by some other means. This might be achieved by surrounding a semiconductor slice with a vapour of the material to be diffused and maintaining the system at a high temperature so that diffusion can take place into the sample. At the temperature and vapour pressure of the experiment, the solute has a solubility in the semiconductor of C_0, say. Let us assume that the surface of the semiconductor achieves this concentration immediately. Then, providing the vapour pressure does not change throughout the experiment (i.e. providing the amount of solute entering the semiconductor is very small compared to the amount initially in the vapour phase) the surface concentration will be maintained at C_0. The boundary conditions are

$$C = C_0 \quad \text{at } x = 0, \text{ all } t$$
$$C = 0 \quad \text{at } t = 0, \, x > 0$$

and the diffusion profile will be of the form

$$C = C_0 \, \text{erfc} \, \frac{x}{2(Dt)^{1/2}} \tag{2.24}$$

where C_0 corresponds to $\frac{1}{2} C'$ in equation (2.23) and the semiconductor occupies the region $0 < x < \infty$.

The same solution can be used for other, related, experimental conditions. Suppose the sample is already doped with the solute, to a concentration C_1, and the external vapour source is used to change the surface concentration to C_2 and maintain it at that level. Equation (2.24) can still be used if we employ the simple expedient of changing the zero of concentration from $C = 0$ to $C = C_1$. The solution becomes

$$C - C_1 = (C_2 - C_1) \, \text{erfc} \, \frac{x}{2(Dt)^{1/2}}. \qquad (2.25)$$

The same equation can be used to cover the case of out-diffusion when a sample which is initially doped to a level C_1 is heated in a vacuum. We assume that the dopant which is diffusing out is pumped away so that no solute vapour collects near the specimen, and put $C_2 = 0$. The solution is

$$C - C_1 = - \, C_1 \, \text{erfc} \, \frac{x}{2(Dt)^{1/2}}$$

i.e.

$$C = C_1 \, \text{erf} \, \frac{x}{2(Dt)^{1/2}}. \qquad (2.26)$$

2.2.5 Rate limitation on surface concentration

In the previous section, we considered cases in which the surface concentration is kept constant by some external phase. The assumption made in defining the initial conditions was that this concentration is assumed 'immediately' at time $t = 0$. This gives rise to the erf solution for out-diffusion and the erfc solution for in-diffusion. In nature, however, nothing happens quite that fast. Solute atoms from the vapour can only join the lattice at a rate which depends on the rate at which vapour atoms impinge on the solid, the probability of such an atom sticking, and the rate at which atoms are re-evaporated. The surface concentration might be considered to go to its equilibrium value 'immediately' if the time constants involved in this kinetic exchange are very small compared to those involved in diffusion. The surface equilibrium is then achieved before diffusion gets underway in the crystal. We will now drop this simplification and consider the implications of a finite time constant for exchange between the surface and the outside world.

The problem is best treated by considering the flow of solute across the interface between the crystal and the vapour at $x = 0$. Define the surface concentration as $C_s(t)$ and let the equilibrium value be C_0. If the rate of evaporation of solute from the solid is proportional to the concentration at the surface, a reasonable assumption, it is given by αC_s, where α is some constant. At equilibrium this becomes αC_0 and is exactly balanced by the rate at which atoms are joining the surface. If $C_s < C_0$, the net rate at which atoms are entering the solid is $\alpha(C_0 - C_s)$. This gives rise to a diffusion flow, J, into the solid, so

$$J = \alpha(C_0 - C_s) \quad \text{at } x = 0 \tag{2.27}$$

or

$$-D\frac{\partial C}{\partial x} = \alpha(C_0 - C_s) \quad \text{at } x = 0. \tag{2.28}$$

It is required to solve the diffusion equation subject to condition (2.28). The solution is given in Carslaw and Jaeger (1959) (although the problem they consider is that of heat flow across the interface rather than flux of atoms):

$$\frac{C}{C_0} = \text{erfc}\,\frac{x}{2(Dt)^{1/2}} - \exp\left(\frac{\alpha x}{D} + \frac{\alpha^2 t}{D}\right)\text{erfc}\left(\frac{x}{2(Dt)^{1/2}} + \alpha\left(\frac{t}{D}\right)^{1/2}\right).$$
$$\tag{2.29}$$

Note that as $t \to \infty$, the second term on the right hand side goes to

$$\exp\left(\frac{\alpha^2 t}{D}\right)\text{erfc}\left(\alpha\left(\frac{t}{D}\right)^{1/2}\right).$$

The exponential tries to take the product to infinity, while the erfc tends to zero. Since the latter function is the stronger of the two, the term goes to zero, so equation (2.29) tends to a simple erfc solution for large diffusion times, as would be expected.

2.3   Finite Difference Methods of Solution

It is usually fairly easy to express a given diffusion problem in terms of Fick's second law and the associated boundary conditions. Unfortunately, there is no guarantee that it can be solved analytically. (It has been shown in the previous section, for instance, how a very minor variation on a standard problem can very quickly lead to a rather complicated solution.) When no analytical solution can be found, it is necessary to resort to the use of numerical methods. Probably the most commonly-used method is that of finite differences; this section deals with the application of the method to the solution of equation (2.7).

For diffusion along the x-axis, we divide the x direction into a number of equal intervals, giving nodes Δx apart. The nodes are numbered so that, for instance, node one might be chosen to be at the surface of a semiconductor slice. Time is similarly divided into intervals Δt. If the concentration at node n at time t is C_n then, following Crank (1956), we describe the concentration at $(t - \Delta t)$ as C_n^-. Using this symbolism, an approximation for the rate of change of concentration at node n can be written:

$$\left(\frac{\partial C}{\partial t}\right)_n = \frac{C_n^+ - C_n}{\Delta t}. \tag{2.30}$$

The right-hand side of equation can be approximated using Taylor's theorem:

$$C_{(n+1)} = C_n + \Delta x \left(\frac{\partial C}{\partial x}\right)_n + \frac{(\Delta x)^2}{2}\left(\frac{\partial^2 C}{\partial x^2}\right)_n + \dots \tag{2.31}$$

$$C_{(n-1)} = C_n - \Delta x \left(\frac{\partial C}{\partial x}\right)_n + \frac{(\Delta x)^2}{2}\left(\frac{\partial^2 C}{\partial x^2}\right)_n - \dots \tag{2.32}$$

where higher terms in the series are ignored. Taking the sum of equations (2.31) and (2.32) gives the approximation

$$\left(\frac{\partial^2 C}{\partial x^2}\right)_n = \frac{C_{(n+1)} - 2C_n + C_{(n-1)}}{(\Delta x)^2}. \tag{2.33}$$

Substituting equations (2.30) and (2.33) into (2.7) gives the recurrence relation

$$C_n^+ = C_n + \frac{D\Delta t}{(\Delta x)^2}\left(C_{(n+1)} - 2C_n + C_{(n-1)}\right). \tag{2.34}$$

So a knowledge of the concentrations at all nodes at time t allows us to calculate the concentrations at $(t + \Delta t)$, and the progress of the diffusion process can be followed. The method has the advantage of simplicity and is much used. Its main restriction is that the constant $D\Delta t/(\Delta x)^2$ is required to be no more than $\frac{1}{2}$, otherwise the computing procedure becomes unstable.

Improvements in the approximations used above are incorporated in the method due to Crank and Nicholson (1947). Consider equation (2.30). This is used in equation (2.34) to provide an estimate of $(\partial C/\partial t)_n$ at time t. Strictly it is the average value between t and $(t + \Delta t)$ and if we must assign it to a specific time, it would be more consistent to choose $(t + \frac{1}{2}\Delta t)$. Let us do this and similarly obtain our estimate of $(\partial^2 C/\partial x^2)_n$ at $(t + \frac{1}{2}\Delta t)$. This can be done by re-writing equation (2.33) at $(t + \Delta t)$

$$\left(\frac{\partial^2 C}{\partial x^2}\right)_n = \frac{C_{(n+1)}^+ - 2C_n^+ + C_{(n-1)}^+}{(\Delta x)^2} \qquad (2.35)$$

and taking the average of equations (2.33) and (2.35). This gives the following

$$C_n^+ = C_n + \frac{D\Delta t}{2(\Delta x)^2} \left[C_{(n+1)} + C_{(n+1)}^+ - 2(C_n + C_n^+) + C_{(n-1)} + C_{(n-1)}^+ \right].$$

$$(2.36)$$

The method converges more rapidly to a solution and is not subject to the restriction $D\Delta t/(\Delta x)^2 < \frac{1}{2}$. It has the disadvantage that it is implicit rather than explicit, i.e. it is not possible to evaluate all values C_n^+ from a knowledge of the C_n quantities. Instead, a series of simultaneous equations must be solved, with the number of equations equal to the number of nodes. This task is usually within the capabilities of a computer, however.

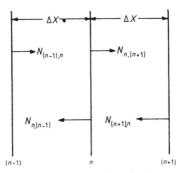

Figure 2.5 Numbers of atoms jumping between adjacent planes in time Δt.

A rather more physical way of looking at the same problem has been suggested by Zahari and Tuck (1982). They modelled the crystal as a series of 'planes', Δx apart, labelled $(n - 1)$, n, $(n + 1)$ as in figure 2.5. If Δx corresponds to the inter-atomic spacing for a crystal, the model is exact; if it is greater (as will almost certainly be the case) then it is an approximation. Initially let the number of diffusant atoms in the planes be $N_{(n-1)}$, N_n, $N_{(n+1)}$ per unit area. The relationships between these quantities and the mean concentrations at the nodes are

$$N_n = C_n \Delta x \qquad (2.37)$$

since the atoms which in the model are concentrated at the nodal planes are, in reality, distributed over a distance Δx.

During a diffusion time Δt, solute atoms jump between planes as shown in the figure. Let us make the assumption that the number which jump is proportional to the number available to jump, i.e.

$$N_{(n-1),n} = kN_{(n-1)} \qquad (2.38)$$

$$N_{n,(n-1)} = N_{n,(n+1)} = kN_n \qquad (2.39)$$

$$N_{(n+1),n} = kN_{(n+1)}. \qquad (2.40)$$

At the end of the step Δt we have

$$N_n^+ = N_n + k(N_{(n+1)} - 2N_n + N_{(n-1)}). \qquad (2.41)$$

Equation (2.37) can be invoked to present this in terms of concentrations:

$$C_n^+ = C_n + k(C_{(n+1)} - 2C_n + C_{(n-1)}) \qquad (2.42)$$

which is the same as equation (2.34) with $k = D\Delta t/(\Delta x)^2$. The restriction $k \leqslant \frac{1}{2}$ now has an obvious physical meaning; if this value is exceeded, we are asking more than 100 percent of the atoms to jump within Δt.

One obvious weakness in this approach is that we are saying that the number available to jump from a node corresponds to the number there at time t. In fact this number will vary during the interval Δt. A better approximation would be to take an average of the number at the beginning and end of the interval:

$$N_{(n-1),n} = \tfrac{1}{2} k(N_{(n-1)} + N_{(n-1)}^+) \qquad (2.43)$$

$$N_{n,(n-1)} = N_{n,(n+1)} = \tfrac{1}{2} k(N_n + N_n^+) \qquad (2.44)$$

$$N_{(n+1),n} = \tfrac{1}{2} k(N_{(n+1)} + N_{(n+1)}^+). \qquad (2.45)$$

Following the argument through in the same way gives us the Crank–Nicholson result of equation (2.36).

So far the modelling approach has achieved little more than reproducing standard results using a physical rather than a mathematical line of argument. The usefulness of the physical method becomes apparent when more complicated diffusion mechanisms are to be modelled. In later chapters examples will be given of this type of modelling.

2.4 Experimental Techniques

In a typical diffusion experiment a dopant is diffused either into or out of a sample and the resulting profile is measured. The form of the profile gives information on the diffusion of the dopant under the experimental conditions employed. Usually the researcher will arrange his experiment so that the boundary conditions correspond to one of the cases which are amenable to mathematical analysis (e.g. the 'thin-film' case or the 'constant surface concentration' case). If he is lucky, the resulting profile will be of the form of the relevant analytical solution (i.e. equation (2.13) or equation (2.24)) and he will be able to assign an unambiguous value of diffusion coefficient, D, to the system under study. If he is not that lucky, the experiment is telling him that he is dealing with a complex diffusion mechanism. This situation is not uncommon in semiconductor work.

A variety of techniques is available for determining the profile, but probably the most direct is the radio-tracer method. The diffusing species used is a radioactive isotope of the dopant of interest and the profile is determined at the end of the experiment by sectioning the semiconductor slice and counting the amount of radioactivity in each section. The technique is accurate and reliable, but is rather time consuming and requires the existence of a suitable isotope. More recently, secondary ion mass spectroscopy (SIMS) has been used for plotting profiles. In this technique, the diffused sample is subjected to sputtering using beams either of oxygen or of caesium. The material sputtered from the surface is analysed by mass spectroscopy and depth profiles are plotted as the semiconductor material is eroded. The method is very accurate, with detection limits of the order of 10^{14} cm^{-3} for some elements, and can plot profiles as shallow as 1 μm or less. This latter characteristic makes it an extremely valuable tool in device work. The main disadvantage is the substantial cost of a commercial SIMS system. Clegg (1982) has recently reviewed the application of the SIMS technique to semiconductor systems.

When a semiconductor contains either shallow donors or shallow acceptors, it is often assumed that all donors (acceptors) are ionised so that, for instance, a profile of ionised donors is the same as an atomic profile of donor atoms. This idea gives rise to a number of electrical techniques for determining profiles which, on the whole, are simpler and quicker than those outlined above. The four point probe method (Smits 1958), for example, measures the conductivity of a semiconductor slice just below the surface. It is valid for diffusions of n-type dopant into p-type substrates or *vice versa*, i.e. a p–n junction must be formed. Four equally spaced probes are applied to the semiconductor surface. The distance between them is usually of the order of 1 mm or less, and they are arranged in a straight line. A current I, of the order of milliamps, is passed between the two outermost probes and the voltage V between the two middle ones is measured. This voltage might typically be in the millivolt range. The relationship between V, I and the conductivity depends on the geometry of the specimen. It has been calculated for most arrangements which are likely to arise in semiconductor work. Using a sectioning technique a conductivity profile can easily be obtained and, if appropriate values of electron (hole) mobility are available, this can be converted into a diffusion profile. The problem here is that mobility is normally a function of concentration and the appropriate data are not always available. If some average value of mobility is used, inaccuracies inevitably occur.

The above problem can be alleviated to some extent by carrying out differential Hall measurements rather than simple conductivity (van der Pauw 1958, Petritz 1958, Buehler 1966). Once again, measurements are taken before and after stripping thin layers from the diffused specimen. Values of mobility and carrier concentration are obtained at each stage and these data can be converted to a profile. The method suffers from the disadvantage of relying on small differences between relatively large measurements.

If a p–n junction is made by diffusing acceptors into a homogeneous n-type sample, then the junction occurs at the depth x_j at which the concentration of acceptors, N_A, is equal to the (constant) concentration of donors, N_D. The depth of the junction is measured in each case and a plot of N_D against x_j gives the diffusion profile for the acceptor. The method obviously requires a large number of starting slices, covering a wide range of homogeneous dopings. Alternatively, we can assume that the profile is of one of the standard forms; in this case, only two substrates are required. Take, for instance, the case of a 'constant surface concentration' diffusion. The expected profile is given by equation (2.24). Suppose we are carrying out identical diffusions of an acceptor into two n-type substrates, of doping N_D and N_D. At the end of the experiment, the two p−n junctions are measured at depths d, d'. Then

$$N_D = C_0 \, \mathrm{erfc}\left(\frac{d}{2(Dt)^{1/2}}\right) \qquad (2.46)$$

$$N_D' = C_0 \, \mathrm{erfc}\left(\frac{d'}{2(Dt)^{1/2}}\right). \qquad (2.47)$$

This is sufficient information to determine C_0 and D. The calculated results are, of course, only as good as the original assumption.

If $N_D \ll C_0$, and if we only want an order of magnitude value for D, an even grosser assumption allows us to determine D from a single diffusion. Since the erfc term must be very small at the junction, we can make use of the approximation that

$$\mathrm{erf}\, z \simeq \frac{1}{\pi^{1/2} z} \exp(-z^2). \qquad (2.48)$$

Substituting this into equation (2.46) and simplifying gives the approximate result

$$d \simeq 2(Dt)^{1/2}\left(\ln \frac{C_0}{N_D}\right)^{1/2}. \qquad (2.49)$$

Since the ratio C_0/N_D enters only by its logarithm, the expression is not very sensitive to its value. Suppose the ratio is 10^4, equation (2.49) becomes

$$d \simeq 6(Dt)^{1/2}. \qquad (2.50)$$

allowing an approximate value of D to be found. All of these p−n techniques, which clearly can be used equally for studying diffusion of donors in p-type material, rely on the accuracy with which the junction depth can be determined. In general, chemical etching or staining methods are used, although techniques also exist which utilise the fact that the Seebeck coefficients for n- and p-type materials are of opposite sign.

The capacitance of a Schottky diode is a function of the applied reverse voltage, V_R. If V_R is plotted against (capacitance)$^{-2}$, a straight line is normally obtained, the slope of which gives the ionised donor (or acceptor) concentration just below the Schottky contact. Once again, a sectioning method can be used to plot the profile. Various clever experimental techniques have been developed in order to facilitate this measurement, including the use of a column of mercury to make the Schottky barrier (Hammer 1969) and a concentrated electrolyte both to form the barrier and to provide a means of controlled anodic dissolution (Ambridge and Faktor 1975).

From the point of view of diffusion studies, the assumption that the concentration of ionised shallow donors (or acceptors) is equal to the concentration of doping atoms represents a major disadvantage in all the electrical techniques. The relationship between the two quantities has been tested for a few systems, using both electrical and non-electrical methods, and in some cases (e.g. the Zn/GaAs system) the numbers tally very well, but in others (e.g. S/GaAs, Zn/InP) large differences can occur. The best practice is to carry out both types of measurement in each investigation; comparison of the two quantities then gives important information on the way in which the atom exists in the semiconductor lattice. This is not often done, however.

2.5 Analysis of Results

2.5.1 'Non-Fickian' diffusion

If an experimentally determined diffusion profile turns out to be of the form of one of the well-known solutions to the diffusion equation, it is a simple matter to assign a value of D to the process under study. If this is not the case, life becomes hard and some other approach must be tried. It is important to understand, however, that a failure to obtain standard profiles does not imply that the basic idea underlying Fick's law is wrong; atoms moving at random in a concentration gradient do obey the law. In many situations, however, the system under study does not involve the simple random motion of atoms and other phenomena are involved, such as reactions between defect species, precipitation etc. These must be taken into account and when this is done, the diffusion profiles, not surprisingly, do not correspond to the standard solutions.

2.5.2   Concentration-dependent diffusion

Under certain special circumstances, an irregular diffusion profile can be dealt with, however. These circumstances, which are fortunately quite common, consist of the following.

(i) The diffusion coefficient is a function of concentration only, i.e. $D = D(C)$. Here it is assumed that the diffusion in question takes place at a single temperature, so that the temperature dependence of D need no be taken into account.

(ii) The boundary conditions can be described in terms of the new variable $\eta = x/t^{1/2}$ (a transformation originally suggested by Boltzmann 1894).

These two conditions form the basis of the Boltzmann–Matano technique (Matano 1932) for analysing diffusion profiles. When they both apply, concentration C can be shown to depend only on the single variable η.

It is convenient to take as example the pair of semi-infinite bars, since we are then looking at the whole solution from $x = -\infty$ to $x = \infty$. The constant surface concentration case is then also included as the positive half of the problem.

Using the nomenclature of sections 2.2.3 and 2.2.4, the boundary conditions are:

$$C = C' (= 2C_0) \quad \text{for } x < 0 \text{ at } t = 0$$
$$C = 0 \qquad\qquad \text{for } x > 0 \text{ at } t = 0$$

and these can be expressed in terms of η

$$C = C' \qquad \eta = -\infty$$
$$C = 0 \qquad \eta = \infty.$$

So here the assumption concerning the boundary conditions is valid. Now consider Fick's second law for a concentration-dependant diffusion coefficient:

$$\frac{\partial C}{\partial t} = \frac{\partial}{\partial x}\left[D(C)\frac{\partial C}{\partial x}\right]. \tag{2.51}$$

Introducing the variable η, we have

$$\frac{\partial C}{\partial x} = \frac{\partial C}{\partial \eta}\frac{\partial \eta}{\partial x} = \frac{1}{t^{1/2}}\frac{\mathrm{d}C}{\mathrm{d}\eta}$$

and, similarly

$$\frac{\partial C}{\partial t} = -\frac{x}{2t^{3/2}}\frac{\mathrm{d}C}{\mathrm{d}\eta}$$

hence

$$\frac{\partial}{\partial x}\left(D\frac{\partial C}{\partial x}\right) = \frac{\partial}{\partial x}\left(\frac{D}{t^{1/2}}\frac{\mathrm{d}C}{\mathrm{d}\eta}\right) = \frac{1}{t}\frac{\mathrm{d}}{\mathrm{d}\eta}\left(D\frac{\mathrm{d}C}{\mathrm{d}\eta}\right).$$

Substituting into equation (2.51),

$$-\frac{\eta}{2}\frac{\mathrm{d}C}{\mathrm{d}\eta} = \frac{\mathrm{d}}{\mathrm{d}\eta}\left(D\frac{\mathrm{d}C}{\mathrm{d}\eta}\right). \tag{2.52}$$

So now both the differential equation and the boundary conditions are expressed in terms of the same variable.

Equation (2.52) can be integrated with respect to η between $C = 0$ and $C = C_1$, where C_1 is any specific value, i.e. $0 < C_1 < C'$:

$$-\frac{1}{2}\int_0^{C_1} \eta\,\mathrm{d}C = \left(D\frac{\mathrm{d}C}{\mathrm{d}\eta}\right)_{C=0}^{C=C_1}. \tag{2.53}$$

The profile under discussion will have been plotted for some specific time, so t can be treated as a constant when x and t are replaced in the above:

$$-\frac{1}{2}\int_0^{C_1} x\,\mathrm{d}C = Dt\left(\frac{\mathrm{d}C}{\mathrm{d}x}\right)_{C=0}^{C=C_1} = Dt\left(\frac{\mathrm{d}C}{\mathrm{d}x}\right)_{C=C_1}. \tag{2.54}$$

where the last equality comes from the fact that $\mathrm{d}C/\mathrm{d}x = 0$ when $C = 0$. Since $\mathrm{d}C/\mathrm{d}x$ is also zero at $C = C'$, we have from equation (2.54) that

$$\int_0^{C'} x\,\mathrm{d}C = 0. \tag{2.55}$$

It follows that for the boundary conditions to be satisfied, the origin from which x is measured must obey equation (2.55),i.e. the equation defines the origin.

From equation (2.54) an expression is found for the diffusion coefficient at $C = C_1$:

$$D(C_1) = -\frac{1}{2t}\left(\frac{dx}{dC}\right)_{C_1} \int_0^{C_1} x \, dC. \qquad (2.56)$$

All of the quantities on the right hand side of equation (2.56) are available from the experiment. The diffusion time t is known and the other quantities are the inverse of the slope at C_1, and the area under the graph.

The procedure is therefore as follows. Taking the curve shown in figure 2.6 as the experimental profile, the zero for the x-axis is first found using equation (2.55); in practical terms this means making the two hatched areas in figure 2.6(a) equal. The value of the diffusion coefficient corresponding to C_1 is then found by measuring the slope at that point and also the shaded area of figure 2.6(b). The values are substituted in equation (2.56) and D_1 is calculated. The process is then repeated for new values of C_1 and the function $D(C)$ is determined.

When the experiment is 'constant C_0' rather than two semi-infinite bars, we have only to solve for positive x. This means that the $x = 0$ position is not decided by an equation but is fixed by the experiment at the surface of the specimen. It is important to remember that one is dealing with only half the solution in this case, and not to get C', the largest concentration measured in the two bar experiment, confused with C_0, the largest measured in the constant C_0 experiment. In fact $C_0 = \frac{1}{2} C'$.

It will be seen that the technique relies heavily on the assumption that it is proper to replace (x, t) by η. The simplest way to check the assumption is to carry out a number of diffusions which are identical except for diffusion time. The times are chosen to cover a wide range and the profiles are plotted as C v. $x/t^{1/2}$. If the assumption is valid, all profiles fall on the same curve. If they do not, then D is probably not a function of C alone. This check is not often done, but when it is, the results sometimes indicate that the assumption is not valid for the diffusion system concerned (see Tuck and Kadhim 1972). The method should therefore be used with caution.

If D is known to be a function of C and it is required to determine the relationship between the two, the isoconcentration method provides the best approach. This technique consists of a series of double experiments. In the first part, dopant is diffused into the semiconductor for a sufficiently long time for the sample to become homogeneously doped to a level C_1, say. If the diffusion source is a vapour surrounding the slice, C_1 will depend on the vapour pressure. A second diffusion is then performed for a shorter time using a radioactive isotope of the dopant at the same vapour pressure. At the end of the diffusion, a profile of the radioactive impurity in the semiconductor is plotted.

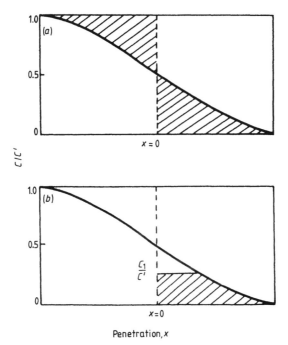

Figure 2.6   Boltzmann–Matano technique.

Because the same amounts of material are used in both experiments, the second one takes place at equilibrium. The semiconductor is surrounded by the correct vapour pressure of dopant and so there is no net exchange of atoms between solid and vapour; the same number of atoms join the vapour from the solid as leave it. Although the semiconductor cannot tell the difference between the atoms joining it from the vapour and those leaving it, we can, since the former are radioactive. The radioactive atoms diffuse in a crystal that is always homogeneously doped to a level C_1 and equation (2.51) becomes

$$\frac{\partial C}{\partial t} = D(C_1)\frac{\partial^2 C}{\partial x^2}. \qquad (2.57)$$

This equation has the standard erfc solution for the experiment described. It gives a value of D for the specified concentration C_1. A whole series of double diffusions can be performed using different values of C_1. Finally a plot of D v. C is obtained. The technique is unambiguous, but rather exhausting.

2.6   Interaction of Defects

It has been noted already that diffusion processes in semiconductors often turn out to be 'non-Fickian', i.e. the profiles do not necessarily correspond to any of the well-known solutions of the diffusion equation. More often than not, the reason for such deviations proves to be interaction between various defects in the semiconductor crystal. Here the term 'defect' is defined very widely; it includes, for instance, crystal vacancies, interstitial atoms, electrons and holes, as well as the diffusing atoms. It is found that the law of mass action, a rule borrowed from physical chemistry, provides a powerful way of looking at these interactions. In this section a simple description will be given of the law, together with some examples of its use. A more detailed description of the application of mass action principles to semiconductor systems is given in Tuck (1974).

2.6.1   Law of mass action

Consider an enclosure containing atoms A and B and suppose that they can interact to form a compound AB

$$A + B \rightleftharpoons AB \qquad (2.58)$$

where the arrows indicate that AB can equally well dissociate into the constituent atoms, so the equation can move right–left as well as left–right. Suppose the concentrations are [A], [B] and [AB] respectively. (The square-bracket notation, indicating concentration of the species, is used extensively in this area of work.)

The frequency with which A atoms meet up with B atoms will depend on the concentrations of both, i.e. the rate at which equation (2.58) moves left–right is given by $k_1[A][B]$, where k_1 is a constant. The frequency with which the AB molecules break up must be proportional to their number, i.e. the equation moves right–left at a rate $k_2[AB]$. If the system is in equilibrium, the rate at which AB is created will balance the rate at which it is destroyed

$$k_1[A][B] = k_2[AB]$$

or,

$$[A][B] = K[AB] \qquad (2.59)$$

where $K = k_2/k_1$ is called an equilibrium constant. In general, K depends only on temperature.

The above example, while instructive, does not cover all eventualities. The substances in question can occur in a number of physical forms, e.g. solid, liquid, gas, as a component of an alloy or the solute in a solution. It is desirable to develop a formulation of the law of mass action which covers all of these possibilities. This is achieved by replacing the concentrations [A], [B], [AB] in equation (2.59), for example, by the more generalised parameters a_A, a_B, a_{AB}, which are called the 'activities' of the substances, so that

$$a_A a_B = K a_{AB}. \qquad (2.60)$$

It is now necessary to define a further term: 'standard state', which provides a reference, permitting any thermodynamic quantity to be quoted relative to its value when the substance is in this state. The standard state is referred to some specified temperature and is usually defined for condensed phases as the pure liquid or solid form of the substance at the specified temperature and one atmosphere pressure. Thus at room temperature the standard state of silicon is pure solid silicon, but the standard state of mercury is pure liquid. For a substance which is normally a gas at the specified temperature, the standard state is simply the gas at that temperature and one atmosphere pressure. Materials are not always in their standard states, of course; a supercooled liquid is not, for example, and neither is a dopant dissolved in a semiconductor.

Now consider a condensed phase A, which in principle may be liquid or solid, and its attendant vapour. Choosing the former phase to be a solid and assuming constant temperature and pressure, the reaction may be written

$$A(\text{solid}) \rightleftharpoons A(\text{gas}, P_A) \tag{2.61}$$

where P_A is the vapour pressure of the gas phase. We assume that equation (2.61) is in equilibrium so that there is no net evaporation of material from the solid phase or deposition from the gas. In the special case of the solid being in its standard state, the vapour pressure is written P_A^0. If the substance is not in its standard state its equilibrium vapour pressure $P_A \neq P_A^0$ and the activity of the substance is defined as

$$a_A = \frac{P_A}{P_A^0} \tag{2.62}$$

from which follows the important result that the activity of a substance in its standard state is unity. If the pure substance at the temperature of interest is a gas, then its standard state is the gas at one atmosphere pressure (i.e. $P_A^0 = 1$ atm). The activity is then numerically the same as the pressure of the gas phase measured in atmospheres. (Note that activity has no units.)

When a substance is in solution it is not in its standard state and its vapour pressure above the solution is generally less than P^0. Suppose the solution is of substance A dissolved in B, and that the mole fraction of A in the solution is M_A. The vapour pressure of A above the solution must obviously vary from zero when $M_A = 0$ and the solution is pure B, to P_A^0 when $M_A = 1$ and the solution is pure A. It is found experimentally that for a limited number of solutions, P_A varies linearly between these extremes i.e.

$$P_A = M_A P_A^0. \tag{2.63}$$

If follows from equation (2.62) that, in this case,

$$a_A = M_A. \tag{2.64}$$

Any solution that obeys equation (2.63) is called ideal. Most solutions are not ideal, but in applying the law of mass action to semiconductor solutions, the approximation of ideality is very often made, and this will normally be the procedure in this book. Equation (2.64) indicates that the activities of components are directly proportional to the concentrations (in units of m^{-3}). Thus when dealing with dopant–dopant or dopant–defect interactions, for instance, simple formulations such as equation (2.59) are appropriate. The one case in which deviation from ideality can cause gross errors is that of 'solutions' of electrons and holes in semiconductors. This point will be taken up in the following section.

2.6.2 Examples of use of law

Let us start with a simple example of an interaction taking place entirely within a crystal. Any crystal will contain vacant lattice sites, concentration [V]. A certain number of these will aggregate to form divacancies. This may occur either because the elastic strain field is lowered by the formation of a divacancy or, more simply, by the chance appearance of vacancies on adjacent sites. In either case, the interaction can be described

$$V + V \rightleftharpoons V_2 \qquad (2.65)$$

and, applying the law of mass action, we have

$$a_V a_V = a_{V_2}. \qquad (2.66)$$

The mole fraction of the vacancies is given to a high degree of accuracy by the ratio of vacancy concentration to site concentration, L, i.e.

$$M_V = \frac{[V]}{L}. \qquad (2.67)$$

Similarly,

$$M_{V_2} = \frac{[V_2]}{\frac{1}{2}L}. \qquad (2.68)$$

We can use equation (2.64) to obtain the following relationship

$$[V]^2 = K[V_2] \qquad (2.69)$$

where K is a constant. Thus the important result is obtained that the concentration of divacancies is proportional to the square of the vacancy concentration. Note the similarity between equations (2.69) and (2.59).

The next example concerns the interaction which takes place at the surface between a solid semiconductor and the surrounding vapour. To fix ideas, let us choose GaAs. At thermodynamic equilibrium, a GaAs sample is surrounded by vapours of gallium and arsenic, with atoms of both elements leaving and joining the surface at the same rate. We assume that arsenic exists in the vapour as the tetramer As_4 and that gallium occurs as the monomer. The equilibrium equation can be written

$$4Ga(vapour) + As_4(vapour) \rightleftharpoons 4GaAs(solid). \qquad (2.70)$$

The activities are as follows:

$$a_{Ga} = \frac{P_{Ga}}{P_{Ga}^0}; \quad a_{As} = \frac{P_{As_4}}{P_{As_4}^0}; \quad a_{GaAs} = 1$$

where P_{Ga}^0 and P_{As}^0 are the vapour pressures over pure gallium and arsenic at the relevant temperature and the activity of a pure substance in its standard state is unity. The law of mass action gives

$$\left(\frac{P_{Ga}}{P_{Ga}^0}\right)^4 \left(\frac{P_{As_4}}{P_{As_4}^0}\right) = K_1(T) \qquad (2.71)$$

and, since the denominators are constants at constant temperature, we have

$$P_{Ga} = K_2(T) P_{As_4}^{-1/4}. \qquad (2.72)$$

This indicates that if, for instance, the experimental conditions are changed so as to increase the arsenic pressure by a factor of 16, the gallium vapour pressure is automatically halved.

In order to cope with the final example, a small creative leap is required. It is quite straightforward to think of a donor, for instance, as a solute dissolved in a semiconductor crystal. It is not stretching the imagination too much to regard a vacancy in a similar light. We now take the view that electrons and holes are also point defects dissolved in the crystal. This idea, which is by no means obvious, proves to be of immense value in considering defect interactions. It can be justified theoretically, but doing this would take us too far from our intended path. The common phenomenon of vacancies becoming charged provides a useful example. Suppose a vacancy can behave as an acceptor under some circumstances, becoming negatively charged. We can then write

$$V^- + h \rightleftharpoons V^\circ \qquad (2.73)$$

where V^0 is an uncharged vacancy and h is a hole. Hence

$$p[V^-] = K[V^0] \qquad (2.74)$$

where p is the hole concentration, which can be replaced by n_i^2/n, where n_i is the intrinsic carrier concentration. Since this latter quantity is a constant at fixed temperature, equation (2.74) can be re-written

$$[V^-] = K'[V^0]n. \qquad (2.75)$$

This gives the interesting result that at a fixed temperature (at which $[V^0]$ is a constant) the concentration of charged vacancies is proportional to the number of free electrons. Thus an increase in the donor concentration will lead to an increase in the charged vacancy concentration.

It has been noted above that equation (2.64) applies only to ideal solutions. In fact, 'solutions' of electrons and holes in semiconductors stray far from ideality if the concentrations become too large. It turns out that these solutions can be considered ideal only if the concentrations are well represented by Boltzmann statistics. In fact, electrons and holes never follow Boltzmann statistics exactly; an accurate description requires the Fermi–Dirac formulation. For low concentrations, however, the data obtained using Boltzmann statistics is indistinguishable from that derived using Fermi–Dirac. The breakdown in this equivalence can occur at surprisingly low concentrations, especially in semiconductors with low effective mass such as gallium arsenide. If the actual concentration of electrons is called n_e and the concentration predicted by Boltzmann statistics is called n_B, then it is n_B that is required in mass-action formulations such as equation (2.75). A sensible compromise is to adjust reality a little by multiplying n_e by some function γ to transform it to n_B, i.e.

$$n_B = \gamma n_e. \qquad (2.76)$$

The term γn_e is then used in the mass-action equation. The quantity γ is called the electron activity coefficient and is a function of electron concentration for those concentrations which are too great to be considered dilute. For dilute concentrations, of course, it is unity. In much the same fashion, an activity coefficient for holes can be defined. Full details on how to calculate γ are given in Tuck (1974).

2.6.3   A warning

The attentive reader will have noticed that section 2.6 has been exclusively concerned with systems at equilibrium, and may well be asking what this has to do with diffusion. It is quite true that if there is a net diffusion flow within a sample it cannot be at equilibrium, by definition. The question to be asked, therefore, is whether mass-action formulations provide results which are good approximations for the non-equilibrium case. As with many such questions, the answer comes down to a comparison of relative time scales. If the diffusion rate is such that very little diffusion occurs in a time scale of 1 ms, say, but the time constant for achieving thermodynamic equilibrium between defects is 1 μs, then it is a very good approximation to assume that equilibrium exists between defect species, even though the sample is not, strictly speaking, at equilibrium. In this case it is quite proper to use the law of mass action in considering results. More often than not it is assumed that the time constants are appropriate for mass action to be used, and in general this practice will be followed in the remaining chapters. The assumption is not often justified, however, and results obtained using it should be considered critically.

2.7   The Built-in Field Effect

When an ionised donor or acceptor diffuses in the presence of an electric field F, Fick's law has to be modified. For an ionised donor, concentration C, the diffusion equation becomes

$$J = - D_D \frac{\partial C}{\partial x} \pm \mu_D FC \qquad (2.77)$$

where D_D and μ_D are the diffusion coefficient and mobility of the ionised donor, related by the Einstein equation

$$\frac{D_D}{\mu_D} = \frac{kT}{q} \qquad (2.78)$$

and a singly ionised donor is assumed. The sign of F in equation (2.77) depends on whether the imposed field is assisting the diffusion or retarding it.

The effect does not necessarily depend on an electric field being imposed from outside, however. When the neutral donor originally enters a lattice, it splits into an ionised donor and an electron, and the two components are free to diffuse separately. The mobility of an electron in a semiconductor lattice, μ_n, is very much greater than μ_D, so the electrons get ahead of the ions, on the average, and a space-charge is set up. The resulting field is such as to help the motion of donors and hinder that of electrons. One might expect, therefore, that the effect of this built-in field is to bring about an increase in diffusion coefficient, and this is indeed so. A fairly simple treatment of the effect for a diffusing donor is given here; a rather more sophisticated version has been presented by Klein and Beale (1966), in which simultaneous diffusion of donors and acceptors is considered.

Consider the diffusion of a donor into an n-type substrate, from the surface. Suppose that the boundary condition at the surface, (i.e. at $x = 0$) is that the concentration of dopant, $C(0)$, is much greater than the background doping of the substrate, C_n. For a sufficiently large value of x, the dopant concentration will, of course, drop below C_n.

According to Boltzmann statistics, the equilibrium concentration of electrons in the conduction band of the semiconductor is given by

$$n = N_c \exp\left(\frac{E_F - E_c}{kT}\right) \tag{2.79}$$

where N_c is the effective density of states in the conduction band, E_F is the Fermi energy and E_c is the energy at the bottom of the conduction band. For a sample at equilibrium, the Fermi level is constant, so it is convenient to take this as the zero of energy. The field at any point inside the crystal is then simply given by the variation of E_c at that point, i.e.

$$qF = \frac{dE_c}{dx}. \tag{2.80}$$

Substitution into equation (2.79) gives, for one dimension

$$F = -\frac{kT}{q}\frac{d(\ln n)}{dx}. \tag{2.81}$$

For heavily-doped semiconductors, equation (2.79) is not appropriate and the Fermi–Dirac expression should be used. Alternatively, the activity coefficient nomenclature can be employed, as described in section 2.6.2, and equation (2.81) becomes

$$F = - \frac{kT}{q} \frac{d(\ln \gamma n)}{dx}.$$ (2.82)

If the semiconductor is non-degenerate, equation (2.81) is correct to a high degree of approximation. It can be re-written

$$F = - \frac{kT}{qn} \frac{dn}{dx}.$$ (2.83)

Close to the surface, the semiconductor is strongly n-type and electrical properties are dominated by the diffusing donor. Since the semiconductor is not degenerate, we can assume that all the donors are ionised. The neutrality condition can therefore be written

$$C = n.$$ (2.84)

The use of this condition may seem a little odd to the reader, since if the semiconductor were completely neutral at all points, there would be no space–charge and hence no field. However it can be shown (Lehovic and Slobodsky 1961) that the magnitude of the space-charge is small compared to C, so while equation (2.84) is not strictly correct, it is still a good approximation.

Substituting equations (2.83) and (2.84) into equation (2.77) gives

$$J = - D_D \frac{\partial C}{\partial x} - \frac{kT}{q} \mu_D \frac{\partial C}{\partial x}$$ (2.85)

where the negative sign has been used from equation (2.77) since the effect of the field is to aid diffusion. Replacing μ_D with D_D from Einstein's relation gives

$$J = - 2D_D \frac{\partial C}{\partial x}.$$ (2.86)

Further into the sample, C drops below C_n. The diffusing dopant then has virtually no effect on the electrical properties in the region and we can write

$$C_n = n$$ (2.87)

and dn/dx is effectively zero. From equation (2.83), the field is zero, so equation (2.77) becomes

$$J = - D_D \frac{\partial C}{\partial x}.$$ (2.88)

Thus the maximum effect of the built-in field for a non-degenerate semiconductor is to double the effective diffusion coefficient.

When the concentration of the diffusing donor becomes sufficiently large to make the semiconductor degenerate, the situation becomes rather more complex. The field is now given by equation (2.82), where the activity coefficient γ is a strong function of concentration. In addition, equation (2.84) becomes invalid, since only a proportion of the concentration of shallow donors is ionised in a degenerate semiconductor. The problem has been considered by Shaw (1974) who showed that in a heavily-doped semiconductor the enhancement of diffusion coefficient of the ionised donors can be greater than the factor of two calculated above. On the other hand, the proportion of donors which are charged decreases sharply as a semiconductor becomes degenerate, so the two effects tend to cancel each other out.

2.8   The External System

Although a diffusion process takes place entirely inside a specimen, we usually cannot ignore what is happening outside. If diffusion is taking place from the surface, for instance, the surface concentration is determined by the external source of diffusant and this concentration has a major influence on the subsequent diffusion. Consider a diffusion in which a dopant C is diffused into a compound semiconductor AB inside a closed ampoule. This is the simplest arrangement to deal with theoretically, since the system under study is well defined; it consists of everything within the ampoule. In principle, the ampoule can contain solids, liquids and gases, each containing up to three elements. The restrictions imposed by the laws of thermodynamics are summed up in the phase rule

$$F = C - P + 2$$ (2.89)

where F is the number of degrees of freedom available to the experimenter, C is the number of components in the system (three in this case, corresponding to the elements A, B, C) and P is the number of phases. Suppose that inside the ampoule there are just three phases: a vapour, a single liquid and a single solid. In this case $F = 2$ from equation (2.89). This means, for instance, that if the temperature of the ampoule has been fixed by putting it into a diffusion furnace and the conditions of the experiment are such as to determine the partial pressure of element B in the ampoule, all other details of the experiment (vapour pressures of elements A, C, compositions of the liquid and solid phases) are fixed and out of the hands of the experimenter.

It is desirable to develop a convenient way of representing the amounts of material within a system. When three elements are involved, as in the case under discussion, the ternary phase diagram is used. To describe the 'closed ampoule' system, we need three parameters, namely temperature plus two to indicate composition. The model used is a prism in which the temperature is plotted on the vertical axis and the base is used to represent composition. It is sensible, for reasons which will soon become apparent, to take the base as an equilateral triangle. A constant-temperature section of the three-dimensional diagram is shown in figure 2.7. The composition axes can be plotted either in terms of weight per cent or mole fraction. In semiconductor work the latter is usually preferred. The three corners of the triangle represent pure A, B, C respectively. Assuming the sides of the triangle to be of unit length, the composition corresponding to a point such as x within the triangle is found as follows. Lines are drawn through x parallel to the sides of the triangle. The mole fractions of the constituents are given by:

$$\text{mole fraction of A} = mC = nB = xs = xp$$
$$\text{mole fraction of B} = rA = sC = xo = xm$$
$$\text{mole fraction of C} = oA = pB = xr = xn$$

where the property of an equilateral triangle of unit side

$$xs + xo + xr = 1$$

has been used. Here the point corresponds to a system containing 0.3A, 0.3B, 0.4C.

All this tells us how to locate any particular composition on the diagram but says nothing about the various phases which would occur inside the ampoule. Such data is found by experiment and, when it is determined, a phase diagram such as that shown in figure

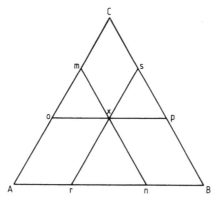

Figure 2.7   Ternary phase diagram.

2.8(*a*) is obtained. The figure, which is rather uncomplicated as phase diagrams go, is similar to that exhibited by some III–V semiconductors when mixed with a group II element (e.g. Ga–P–Zn). The line bcde in figure 2.8(*b*) is called a solidus; all material within this curve is solid AB containing some C in solution. The extent of this region has been greatly enlarged in the figure, for the sake of clarity. The distance b – e is a measure of the extent of non-stoichiometry which is permitted for the compound AB, b being the point of maximum A composition and e being the point of maximum B composition. For many compounds, including the III–V compounds, this permitted range of variation is very small indeed.

The line ahgf is called a liquidus; all compositions outside this line are liquids, containing all three elements. That part of the diagram lying between the solidus and the liquidus corresponds to systems in which solid and liquid phases occur in equilibrium. The relative amounts of the two phases are found by determining the tie line which joins together the solid composition on the solidus with the liquid composition with which it is in equilibrium on the liquidus. Two such tie lines are shown in figure 2.8(*b*). Suppose the system inside the ampoule is known to correspond to the point x in figure 2.8(*b*). A 'lever' law similar to that already described applies:

mole fraction of solid phase at equilibrium = xh/ch
mole fraction of liquid phase at equilibrium = xc/ch.

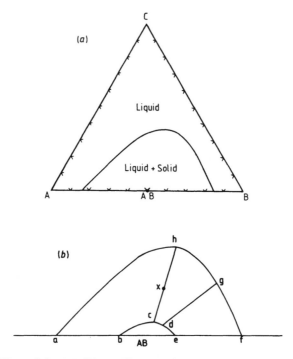

Figure 2.8 (*a*) Phase diagram for A–B–C system at temperature *T*. (*b*) Enlarged view of central portion of phase diagram; the size of the solid region is greatly exaggerated.

It should be noted that we have made no specification concerning the vapour pressure in the ampoule. In many metallurgical systems, the practice is to plot phase diagrams for a constant over-pressure of one atmosphere. Here it is preferable to allow the partial pressures to be determined by the condensed phases, i.e. the pressures will be different at different points within the triangle, even though the temperature is constant. To give a complete specification for the system, therefore, further information is required giving the vapour pressures P_A, P_B, P_C as functions of the composition of the condensed phases. Unfortunately this information is not often available; one notable exception is the Ga–As–Zn system, which will be discussed in some detail in Chapter 4.

2.9   Assessment of Published Data

Given the enormous number of possible diffusion systems, all
published experimental data is of value, providing it has been taken
with proper care. At the very least, it gives the reader some idea of
the order of magnitude of diffusion penetration under a given set of
conditions. If the data is being used to draw general conclusions
about diffusion, or to support a proposed diffusion mechanism,
however, then rather stringent criteria are required. There are many
published collections of diffusion data and, on the whole, they tend
not to employ these criteria; it is more usual to take each paper at
its face value and simply pass on conclusions. For the worker who
is seriously interested in diffusion in semiconductors, therefore,
there is no alternative to a critical reading of the original papers.
The following points are intended to help in that critical reading.

(1) If it is assumed *a priori* that the system under test obeys
Fick's law, then values of diffusion coefficient and activation
energy can be obtained from a very small number of experimental
results. P−n junction methods often use this type of approach.
This really amounts to assuming the answer before the experiments
are done. Many semiconductor systems (perhaps the majority) do
not obey Fick's law and it is therefore necessary to carry out
experiments for each individual system to establish whether the law
is followed.

(2) When diffusion profiles are plotted, the conclusion is often
drawn that a curve is a Gaussian or some other standard function
on quite inadequate evidence. Three or four points can be fitted
(more or less) to any function the experimenter fancies. An
experimental profile should have as many points as possible before
fitting is attempted and, ideally, should cover three orders of
magnitude of diffusant concentration.

(3) A good deal of diffusion work in semiconductors is performed using electrical techniques, in which concentrations of
electrons (or holes) rather than concentrations of atoms are plotted. The assumption is made, often implicitly, that the two types of
concentration are always the same. Until it is confirmed experimentally, this is no more than a guess.

(4) Similarly, the Boltzmann−Matano method is frequently
used without any apparent awareness of the fact that it is only valid

for certain systems (see section 2.5.2). It is necessary to carry out subsidiary experiments to discover whether or not it is valid for the particular system under study.

(5) Papers which suggest that a III–V system can be described by a single value of diffusion coefficient at some temperature should be scrutinised very carefully. For many elements, diffusion in III–V semiconductors is characterised by a coefficient which is a function of concentration and, in most systems, diffusion is dependant on stoichiometry which, in turn depends on the ambient vapour pressures. An example is given in Chapter 4 of a system in which D varies by five orders of magnitude at a single temperature.

If we were to restrict ourselves to a consideration only of the experimental studies which have complied with all of the above guidelines, this would be a rather short book. The policy in the following chapters is to give the best evidence available; the reader may detect a note of scepticism from time to time.

Chapter 3

Diffusion of Shallow Donors, including Group IV

In this chapter we consider diffusion of those elements which give rise to donor states close to the conduction band in III−V semiconductors. These include the group VI elements sulphur, selenium and tellurium which sit on the group V sites and therefore have an electron to spare. Also included are the group IV elements tin, silicon and germanium which are able to occupy either the group III site, in which case donor action would be anticipated, or the group V site, when they would be expected to act as acceptors. They are included in this chapter because in practice they usually behave as donors over-all. In general the donors are characterised by low diffusion coefficients although some recent work on InP, to be described in a later section, suggests that this may not always be the case.

3.1 Sulphur

The diffusion of sulphur in GaAs has been studied by a number of workers. Early work was largely concerned with the sulphide layer which can form on the surface of the semiconductor during a diffusion experiment in which sulphur is diffused from the vapour phase. Various techniques were advanced to eliminate this undesirable coating, including covering the GaAs with a thin layer of silicon oxide before diffusion (Yeh 1964) and using a compound source (Frieser 1965). It was generally agreed, however, that if the amount of sulphur in a diffusion ampoule is no more than a few $\mu g\ cm^{-3}$, no sulphide forms. In this early work, the diffusion was mainly studied either by measuring p–n junction depths (Vieland 1961, Frieser 1965, Yeh 1964) or by taking sheet resistivity measurements (Kendall 1968). Neither of these techniques gives detailed information on diffusion properties, although Vieland (1961) did hazard the opinion that nearest-neighbour gallium vacancies might be involved.

A few investigations have been carried out using radio-tracers to ascertain the form of the diffusion profiles. The first of these (Goldstein 1961) was somewhat marred by the subsequent discovery (reported as a private communication in Kendall 1968) that relatively large amounts of GaAs slices had evaporated during the course of the experiment. In a later investigation, Young and Pearson (1970) studied the diffusion of sulphur in GaAs in the range 1000–1300 °C using radioactive ^{35}S. They also varied the amount of sulphur in the diffusion ampoule so that at 1130 °C, for instance, diffusion profiles were obtained with surface concentrations, C_0, in the range $3 \times 10^{18} - 5 \times 10^{20}$ cm^{-3}. The profiles were not the erfc curves that might have been expected from the conditions of the experiment; substantial deviations from this function were observed, the deviations being largest for the greatest values of C_0. Young and Pearson circumvented this problem by drawing 'best fit' erfc curves through the experimental points. In this way they were able to assign a diffusion coefficient D to each of their curves. This must be seen as a rather dubious procedure, and their values of D must be used with some caution.

Young and Pearson also carried out a series of experiments in which the amount of sulphur in the ampoule was kept constant, but the arsenic pressure was made to vary by putting small amounts of metallic arsenic in the ampoules. Two such series were performed; one at 1000 °C and one at 1130 °C. At both temperatures it can be assumed that the arsenic species As$_4$ predominates over As$_2$. The variation of sulphur diffusion coefficient with As$_4$ vapour pressure is shown in figure 3.1. D is seen to increase approximately as the square root of the arsenic pressure up to about 0.5 atm, after which it saturates. Young and Pearson sought to explain this result by proposing that sulphur occurs in the GaAs lattice in two forms: the usual donor S_{As}^{+} which is assumed to be immobile and the mobile complex ($V_{Ga}S_{As}^{+}V_{Ga}$). According to this model, all diffusion takes place via the complex. The complex is shown diagramatically in figure 3.2. It is fairly clear that the absence of two gallium atoms makes the exchange of the sulphur and arsenic atoms much easier. Further progress requires substantial movement of gallium vacancies, however.

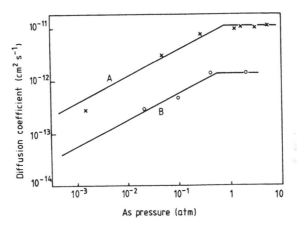

Figure 3.1 Dependence of sulphur diffusion coefficient in GaAs on the ambient arsenic vapour pressure (from Young and Pearson 1970). Reproduced by permission of *Pergamon Journals Ltd*.

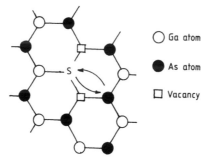

Figure 3.2 Sulphur diffusion in GaAs with the assistance of two gallium vacancies.

Before developing the model further, it is necessary to establish a relationship between the concentration of divacancies and the ambient arsenic pressure. Consider the exchange between gallium atoms in the vapour and those at the surface of the crystal:

$$Ga_{vap} + V_{Ga} \rightleftharpoons Ga_{Ga} \tag{3.1}$$

where Ga_{vap} is a gallium atom in the vapour, V_{Ga} is a vacancy at the surface, and Ga_{Ga} is a gallium atom on a gallium site. The mass-action formulation for equation (3.1) is

$$P_{Ga}[V_{Ga}] = K_1 \tag{3.2}$$

where the square-bracket notation has been used to indicate concentration, and it is noted that the activity of gallium atoms on gallium sites is unity; K_1 is a constant. The relationship between gallium and arsenic partial pressures was discussed in section 2.6.2. Using equation (2.72), we have

$$[V_{Ga}] = K_2 P_{As_4}^{1/4}. \tag{3.3}$$

Similarly, using equation (2.69) with slightly changed nomenclature,

$$[V_{Ga}V_{Ga}] = K_3[V_{Ga}]^2 \tag{3.4}$$

where $[V_{Ga}V_{Ga}]$ is the concentration of gallium divacancies. Equations (3.3) and (3.4) combine to give the desired result:

$$[V_{Ga}V_{Ga}] = K_4 P_{As_4}^{1/2}. \tag{3.5}$$

We can now return to the argument of Young and Pearson. The equilibrium between the mobile and immobile species is given by

$$(V_{Ga}V_{Ga}) + S_{As}^+ \rightleftharpoons (V_{Ga}S_{As}^+V_{Ga}). \tag{3.6}$$

It helps to simplify the notation a little here if we call the donor complex C^+. The mass action equation for equation (3.6) is

$$[V_{Ga}V_{Ga}][S_{As}^+] = K_5[C^+]. \tag{3.7}$$

Since the sulphur either exists in one form or the other, we also have

$$[S] = [S_{As}^+] + [C^+] \tag{3.8}$$

where $[S]$ is the total concentration of sulphur. Combining equations (3.5, 3.7) and (3.8) gives

$$[C^+] = \left(\frac{K_6 P_{As_4}^{1/2}}{1 + K_6 P_{As_4}^{1/2}} \right) [S] \tag{3.9}$$

and

$$[S_{As}^+] = \left(\frac{1}{1 + K_6 P_{As_4}^{1/2}}\right)[S].\qquad(3.10)$$

Writing Fick's law for the total sulphur concentration of equation (3.8),

$$\frac{\partial[S]}{\partial t} = \frac{\partial}{\partial x}\left(D_c\frac{\partial[C^+]}{\partial x}\right)\qquad(3.11)$$

where D_c is the diffusion coefficient of the complex and there is no term for the diffusion of S_{As}^+ since it is immobile. Equation (3.11) can be written

$$\frac{\partial[S]}{\partial t} = \frac{\partial}{\partial x}\left(D_c\frac{\partial[C^+]}{\partial[S]}\frac{\partial[S]}{\partial x}\right)\qquad(3.12)$$

i.e. the sulphur diffuses with an effective coefficient

$$D = D_c\frac{\partial[C^+]}{\partial[S]}\qquad(3.13)$$

or,

$$D = D_c\left(\frac{K_6 P_{As_4}^{1/2}}{1 + K_6 P_{As_4}^{1/2}}\right).\qquad(3.14)$$

Young and Pearson point out that equation (3.14) is in agreement with figure 3.1. At low arsenic pressure, the concentration of gallium vacancies is low. The concentration of the mobile species is therefore small, as is the effective diffusion coefficient. As the arsenic pressure increases, the concentration of C^+ goes up and there is a corresponding increase in the effective diffusion coefficient. Finally, at some high arsenic pressure, all of the sulphur is in the form of the mobile complex and the diffusion coefficient saturates.

Before leaving this work, one other important result must be mentioned. Young and Pearson also used a plasma reflection technique to measure the electron concentration close to the surface of a diffused specimen. By etching away thin layers of GaAs between measurements, they were able to plot an electron concentration profile which could be compared to the diffusion profile of sulphur atoms. Both profiles are shown in figure 3.3 for a sample diffused at 900 °C. It can be seen that for most of the profile, the atomic concentration exceeds that of electrons. This experiment gives a good example of the dangers inherent in using electrical techniques to measure diffusion profiles. It also presents some problems to the model of Young and Pearson, since equation (3.8) implies that all of the sulphur is charged.

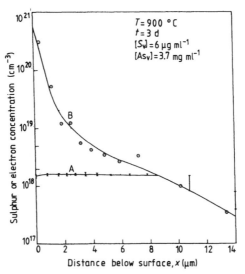

Figure 3.3 Comparison of the electron (line A) and sulphur (curve B) concentrations in sulphur-diffused GaAs (from Young and Pearson 1970). Reproduced by permission of *Pergamon Journals Ltd.*

Diffusion profiles of ^{35}S in GaAs, taken at 1000 °C and 1120 °C were also presented by Tuck and Powell (1981). These curves, shown in figure 3.4, looked very much like those of Young and Pearson (1970). The work was used by Zahari and Tuck (1982) to develop a model for the S/GaAs system which was rather different to the one described above. They pointed out that in this system, the sulphur concentration close to the surface can become comparable to the concentration of arsenic vacancies in the crystal. If the sulphur is diffusing via arsenic vacancies, then a reduction of $[V_{As}]$ below the equilibrium value might be expected. The model which they used is essentially a development of the one illustrated in figure 2.5. In the model described in section 2.3 the assumption is made

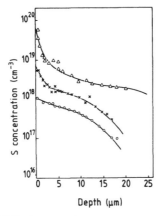

Figure 3.4 Diffusion profiles of sulphur in GaAs at 1120 °C for 1 h. The three curves correspond to three different ambient sulphur pressures (from Tuck and Powell 1981).

that the number of diffusing atoms which jump from node n to node $(n + 1)$ during a diffusion time Δt is proportional to the concentration at n at the beginning of the time-step, C_n (see equations (2.37), (2.38)). This formulation is quite satisfactory providing the concentration of vacancies in the crystal is not substantially affected by the diffusing impurities. If this is not the case, however, allowance must be made for self-diffusion. It is then necessary to make the more precise statement that the number of atoms jumping from plane n to plane $(n + 1)$ is proportional to the number of atoms available to jump on plane n and also to the number of spaces (i.e. vacancies) on plane $(n + 1)$. Equations (2.38) to (2.40) become

$$N_{(n-1),n} = k_1 C_{(n-1)} V_n \tag{3.15}$$

$$N_{n,(n-1)} = k_1 C_n V_{(n-1)} \tag{3.16}$$

$$N_{n,(n+1)} = k_1 C_n V_{(n+1)} \tag{3.17}$$

$$N_{(n+1),n} = k_1 C_{(n+1)} V_n. \tag{3.18}$$

V_n is the concentration of vacancies on plane n and the thermal equilibrium concentration of vacancies in the crystal is V'. The constant is given by

$$k_1 = \frac{D \Delta t}{(\Delta x)^2 V^1}$$

by analogy with the constant defined in section 2.3; if $V_n = V'$ for all n, then equations (2.38)–(2.40) become identical to equations (3.15)–(3.18). The impurity concentration at node n after a diffusion step Δt now becomes.

$$C_n^+ = C_n + k_1 \left(C_{(n-1)} V_n + C_{(n+1)} V_n - C_n V_{(n-1)} - C_n V_{(n+1)} \right). \quad (3.19)$$

The self-diffusion of host atoms is taken into account in the same way. If H_n is the number of host atoms at node n at the start of an iteration of duration Δt, then

$$H_n^+ = H_n + k_2 (H_{(n-1)} V_n + H_{(n+1)} V_n - H_n V_{(n-1)} - H_n V_{(n+1)}) \quad (3.20)$$

where

$$k_2 = \frac{D_H \Delta t}{(\Delta x)^2 V'}$$

and D_H is the self-diffusion coefficient for host atoms.

The number of arsenic sites in the crystal is a fixed quantity and each one is occupied by a host atom, an impurity atom or a vacancy. Thus if there are L lattice sites per unit volume,

$$L = H_n + C_n + V_n \quad (3.21)$$

for all values of n.

The computation procedure used is as follows. At time $t = 0$, the nodes are set up with

$$\begin{aligned} C_1 &= C_0 & V_1 &= V' \\ C_n &= 0 & V_n &= V' & (n > 1). \end{aligned}$$

The values of H_n are determined from equation (3.21). The quantities at node 1 are fixed throughout the diffusion by the program. For $t > 0$ the concentrations of impurity and host atoms are found at each node using equations (3.19) and (3.20) and the vacancy concentrations are found, after each iteration, from equation (3.21). For the node with the highest value of n, a reflection condition is applied.

Using the above technique, Zahari and Tuck (1982) produced the profiles shown in figure 3.5, using values appropriate to the S/GaAs system. It can be seen that they demonstrate all of the features shown by the experimental curves of figure 3.4. Further modelling work was carried out in which the effect of varying the external arsenic vapour pressure in the S/GaAs system was investigated.

(V' in the above calculation corresponds to the arsenic vacancy concentration in this case which, in turn, depends on the external arsenic vapour pressure.) The calculated profiles showed the same qualitative behaviour as that demonstrated by the experimental results.

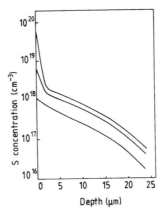

Figure 3.5   Calculated profiles for sulphur in GaAs (from Zahari and Tuck 1982).

It has been noted in Chapter 2 that a knowledge of the appropriate phase diagram can be of great assistance in understanding diffusion processes. A good example of this was reported by Matino (1974), who demonstrated that an intelligent use of the diagram permits highly reproducible diffusions of sulphur in GaAs. The same technique has since been employed by Prince *et al* (1986). The diffusion temperature used by Matino was 910 °C. His proposed phase diagram for Ga–As–S at this temperature is shown in figure 3.6. He derived the diagram from a knowledge of the melting-points of all the solids involved and by assuming that it is of the same general form as the Ga–As–Te diagram, which had been determined by Panish (1967). The regions of interest to Matino are those marked A and B on figure 3.6. Any system described by a point in region A contains three solids in equilibrium, namely GaAs, GaS and Ga_2S_3, together with an ambient vapour phase. The relative amounts of the three solids depend, of course, on where the point is within the region. We therefore have a system of four phases (three solids and a vapour) and three components. Applying the phase rule (equation 2.89), the number of degrees of freedom, F, is one. This has already been used, however, in deciding that the temperature is to be 910 °C. It follows that the composition of the vapour is the same for all points within the region. Region A is called invariant and all diffusions carried out on systems within it should have the same result. This is important as far as diffusion procedures are concerned, since it is not possible in practice to produce consecutive diffusion sources of identical composition. Using this method, it does not matter what the precise composition is, so long as it is within region A. Region B is similarly invariant, with three condensed phases, GaAs, Ga_2S_3 and X (a liquid), in equilibrium. Confining the diffusion source to this region gives similarly reproducible results (although it should be understood that the reproducible result obtained using region B is different to that obtained using region A, since the vapour pressures are different in the two cases). It is perhaps worth noting, however, that these fairly clear statements concerning invariant systems become slightly more complicated when the diffusion of zinc in GaP is considered in the next chapter.

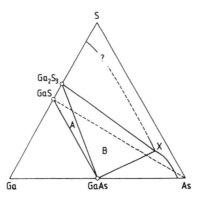

Figure 3.6 Proposed form of Ga–As–S ternary phase diagram (from Matino 1974). Reproduced by permission of *Pergamon Journals Ltd.*

In the previous chapter it was mentioned that when a diffusion process is simple, the variation of diffusion coefficient with temperature often follows the rule

$$D = D_0 \exp(-Q/kT).$$

Diffusion processes in the III–Vs are frequently not simple, however, and it is somewhat risky to assume that the above equation tells the whole story. The point is nicely made by Matino's results. He carried out two series of diffusions in the temperature range 750 °C–910 °C, one in region A and one in region B. In all cases he fitted values of D to the diffusion data, which were obtained using p–n junction depth measurements. By plotting (log D) against T^{-1}, he obtained D_0 and Q. The values calculated for region A diffusions, however, were quite different to those found from region B.

The results obtained from Matino's experiments and from those of other workers are given in table 3.1. The large discrepancies shown do not imply poor technique on the part of the experimenters; the results were, no doubt, correct for the particular conditions of their experiments. For a compound semiconductor, however, there is, in general, no unique value of D for a single temperature.

Table 3.1

$D_0(\mathrm{cm}^2\,\mathrm{s}^{-1})$	$Q(\mathrm{eV})$	Temp. range (°C)	Reference
1.2×10^{-4}	1.80	900–1050	Vieland (1961)
2.6×10^{-5}	1.86	900–1100	Frieser (1965)
1.6×10^{-5}	1.63	700–1000	Kendall (1968)
1.85×10^{-2}	2.6	900–1200	Young & Pearson (1970)
3.78×10^{6}	9.35	810– 910	Matino (1974) region A
10.9	2.95	750– 910	Matino (1974) region B
0.27	3.0	800– 900	Prince *et al* (1986)

Young and Pearson (1970) studied the diffusion of sulphur in GaP over the temperature range 1100 °C–1300 °C and it is interesting that the results were quite different from their S/GaAs work, described above. While the diffusion coefficient of sulphur was sharply dependent on the arsenic vapour pressure over GaAs (see figure 3.1), it was independent of the phosphorus pressure for the GaP system. This result eliminates the simplest possible mechanism, of the sulphur atom jumping from one phosphorus site to a vacancy on the next, since the concentration of phosphorus vacancies is dependent on the phosphorus pressure. They proposed that the diffusion mechanism involves divacancies. Before their argument can be understood, it is first necessary to establish an important result concerning the vacancy concentrations.

Consider the exchange of phosphorus atoms between the vapour and the GaP crystal:

$$(\tfrac{1}{2}P_2)_{vap} + V_P \rightleftharpoons P_P \qquad (3.22)$$

where it is assumed that the species in the vapour is P_2. The mass action equation is

$$P_{P_2}^{1/2}[V_P] = K_1. \qquad (3.23)$$

Equation (3.2) will stand just as well for GaP as for GaAs, so can be used again

$$P_{Ga}[V_{Ga}] = K_2. \qquad (3.24)$$

The relationship between the two partial pressures has been derived for GaAs in chapter 2. Re-writing equation (2.72) for GaP gives

$$P_{Ga}P_{P_2}^{1/2} = K_3 \qquad (3.25)$$

Multiplying equations (3.23) and (3.24), and substituting from equation (3.25) gives

$$[V_{Ga}]\,[V_P] = \frac{K_1 K_2}{K_3}. \qquad (3.26)$$

This result is quite general for III–V semiconductors; at constant temperature, the product of the group III and group V vacancy concentrations is a constant. Thus if the vapour pressures of the group III and group V components over the crystal are changed, both types of vacancy will change concentration, but in such a way as to maintain the same product. There is an obvious analogy here with the product of the electron and hole concentrations in a semiconductor.

Let us now return to the hypothesis of Young and Pearson. They proposed that sulphur diffuses in GaP by means of the divacancy $(V_{Ga}V_P)$. According to this mechanism, a sulphur atom on a phosphorus site jumps to the phosphorus vacancy end of the divacancy. Although the basic jump is the same as for a 'simple' process, the presence of the intervening gallium vacancy means that less distortion of the lattice is required. In this model, D will be proportional to the concentration of divacancies. Consider the formation of a divacancy from two vacancies:

$$V_{Ga} + V_P \rightleftharpoons (V_{Ga}V_P).$$ (3.27)

Mass action gives

$$[V_{Ga}] \, [V_P] = K[V_{Ga}V_P].$$ (3.28)

It follows from equation (3.26) that the concentration of divacancies is a constant, irrespective of the phosphorus vapour pressure. This is in agreement with the result of Young and Pearson, described above.

A rather different approach was taken by Dutt *et al* (1984) when they studied the diffusion of sulphur in InP. Following their earlier work (Chin *et al* 1983a, 1983b), they started with InP which had been sulphur-doped during growth by the Czochralski method. The experimental technique used was to anneal slices of the material, usually for 30 minutes, at temperatures between $350\,°C$ and $650\,°C$ and to look for signs of out-diffusion. A single SIMS profile is shown in their paper, taken after an anneal for 5 minutes at $550\,°C$. The profile shows a dip at $20\,\mu m$ to a concentration of about 90 percent of the original doping. Dutt *et al* attributed this dip to out-diffusion although, interestingly, the surface concentration is not reduced below the bulk value. The diagram is reproduced in figure 3.7. It should be noted that a 'simple' diffusion process would give a profile of the form of equation (2.26). This solution of Flick's law bears no similarity to the curve of figure 3.7. It is not easy, therefore, to derive a value of D from the result.

After an anneal, the sample was cleaved and the surface perpendicular to the faces of the slice was stained with a ferricyanide etch. In all cases, a line was observed parallel to the surface at a depth which ranged from a few microns to more than a hundred, depending on the conditions of the anneal. The depth of the etched line for the sample of figure 3.7 is indicated on the diagram. Dutt *et al* assumed from this result that the depth of the etched line gave a good indication of the position of the dip for all samples and calculated a value for the diffusion coefficient using the somewhat arbitrary relation $D = x_j^2/4t$, where x_j is the depth of the line. An Arrhenius plot of D against $1/T$ was produced and a straight line through the points in the temperature range $350\,°C–500\,°C$ gave a relation of the form of equation (2.2) with $Q = 2.3\,eV$ and

$D_0 = 8.2 \times 10^6$ cm^2 s^{-1}. Points in the range 500 °C–650 °C did not lie on the line, however. The calculated values of D were large compared to the corresponding diffusion rates reported by other workers for sulphur in GaAs. (The results cannot be compared at the same temperature, of course, since the melting-points of the two semiconductors are different. A rule of thumb for comparing diffusion coefficients of the same diffusant in different semiconductors is to make the comparison at temperatures which are the same fraction of the melting temperatures, expressed in K.) In view of the approximations and assumptions made in deriving the above figures, they should, perhaps, be accepted only as semi-quantitative results. There seems no doubt, however, that the phenomenon reported by Dutt *et al* is of interest and warrants further investigation.

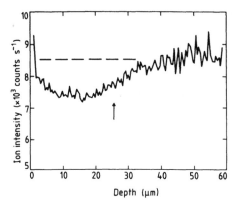

Figure 3.7 SIMS profile of sulphur-doped sample after 5 min anneal at 550 °C (from Dutt *et al* 1984).

The diffusion of sulphur in InAs was studied in an early publication of Schillmann (1962). Using the p–n junction method, he obtained values of D over the temperature range 600 °C–800 °C. He was able to fit them to an expression of the form of equation (2.2), with $D_0 = 6.78$ cm^2 s^{-1} and $Q = 2.2$ eV. In this work he identified the position of the p–n junction by measuring thermo-electric voltages. Information on diffusion in InSb was given by Rekalova *et al* (1971a) who used as source a (In + 1% S) mixture alloyed to the surface prior to the diffusion anneal. The anneals were performed over the temperature range 360 °C–500 °C and capacitance techniques were employed to determine effective diffusion coefficients. An activation energy of 1.4 eV was reported for this process.

3.2   Selenium and Tellurium

Most investigations of the diffusion of selenium and tellurium in III–V compounds were carried out some years ago. When the diffusion source is some phase outside the semiconductor, there is a possibility of forming Ga_2Se_3 or Ga_2Te_3 on the surface and it is likely that this did happen in some of the reported investigations. Fane and Gose (1963) studied selenium diffusion in GaAs from the vapour phase at 1100 °C. They obtained profiles using both the p–n junction and radio-tracer techniques. The latter set of measurements showed a profile shape which did not follow an erfc form, but resembled the sulphur profiles described in section 3.1. They were not able, therefore, to assign an unambiguous value of D to any of these profiles. It is worth noting, however, that a plot of p–n junction depth against (diffusion time)$^{1/2}$ gave a good straight line. This type of plot is often used as a test to establish that the diffusion process is Fickian.

More recently, the Se/GaAs system was studied by Khludkov and Lavrishchev (1976), also using the p–n junction method. Their plot of junction depth v. $t^{1/2}$ did not give a straight line but they nevertheless assumed that the diffusion profiles must be erfc curves and calculated effective values of diffusion coefficient, D_{eff}. The values of D_{eff} were determined by fitting an erfc to two points of the concentration profile; a point at the surface and a point corresponding to the p–n junction. They assumed a surface concentration of 1×10^{20} cm^{-3}, although it is not clear where this number came from. Experiments were carried out over a range of ambient arsenic pressures and D_{eff} was found to be almost independent of this parameter. Electrical measurements were made and the conclusion was drawn that some of the selenium atoms were inactive at high concentrations. This could be due to the formation of complexes with point defects or to the precipitation of Ga_2S_3 inside the slice. Khludkov and Lavrishchev proposed a diffusion mechanism whereby the selenium atom moves via divacancies ($V_{Ga}V_{As}$).

This is certainly consistent with their observation that D_{eff} is independent of the arsenic over-pressure since, as shown in the previous section, a simple mass-action formulation indicates that $[V_{Ga}V_{As}]$ is a constant. The same group (Karelina *et al* 1974) produced a very similar paper on the diffusion of tellurium in GaAs, again using the p–n junction method, and measuring D_{eff} for a range of ambient arsenic pressures, between 1000 °C and 1150 °C. They came to the conclusion that the principal diffusion mechanism in this case operates via arsenic vacancies or divacancies.

The diffusion of both selenium and tellurium in p-type InSb was studied by Rekalova *et al* (1969, 1971b). The technique was essentially the same as the one they used for sulphur, described in section 3.1. A mixture consisting either of (In + 0.5% Te) or (In + 1% Se) was alloyed to the InSb top surface, in a hydrogen atmosphere. This gave rise to a thin recrystallised layer on the surface, heavily doped with the group VI component. The dopant was then driven into the sample at diffusion anneal temperatures in the range 350 °C–500 °C. The progress of the p–n junction was monitored using capacitance techniques. The main conclusion drawn by these workers was that selenium diffuses rather more slowly than tellurium over the temperature range studied. The diffusion of both elements was also studied by Schillman (1962) for InAs but in this material the reverse seems to be the case. For the temperature range 600 °C–900 °C selenium was reported as diffusing faster than tellurium.

The work of Dutt *et al* (1984) on the out-diffusion of sulphur from InP has been described in section 3.1. The same group (Chin *et al* 1983a) produced a very similar result for selenium. The large diffusion coefficient for selenium in InP implied by their results is surprising in view of the small coefficients found for this element in the other III–V semiconductors. Unfortunately, there is no reliable data on in-diffusion of selenium in InP with which this work can be compared.

3.3 Tin

The major electronic effect of tin in the III–Vs is for the atom to go onto the group III site and act as a shallow donor. There is a good deal of experimental evidence to suggest that this is not the whole story, however, and it seems likely that tin can be incorporated in more than one way. In an extensive series of experiments on GaAs, for instance, Panish (1973) grew tin-doped material from the liquid phase and carried out electrical measurements. He came to the conclusion that tin exists in the GaAs lattice in three distinct forms: a positively-charged donor on a gallum site, Sn_{Ga}^+, a negative acceptor complex consisting of a tin atom on an arsenic site plus two gallium vacancies, $(V_{Ga}Sn_{As}^-V_{Ga})$, and a neutral species which might be either another complex or a precipitate. Similar, although not identical, results and conclusions were published by Nishizawa *et al* (1973) and Bolkhovityanov and Bolkhovityanova (1975). Early work by Goldstein and Keller (1961) and by Fane and Goss (1963) indicated that tin diffusion in GaAs is a very slow process compared to that of acceptors in the same material. More recently it has been established that tin diffusions can be carried out from doped films of SiO_2 deposited on the GaAs surface (Gibbon and Ketchow 1971, Yamazaki *et al* 1975, Arnold *et al* 1984).

The largest body of published experimentation is that due to Tuck and Badawi (1978) who diffused radioactive tin into n-type GaAs from the vapour phase over the temperature range $850\,^{\circ}C{-}1100\,^{\circ}C$, and plotted the resulting diffusion profiles. Initially they experienced difficulty in obtaining reproducible results. Investigation revealed that this was due to loss of material from the GaAs slice during the course of the experiment. Because of the slow diffusion of tin into GaAs, the profiles extended only about 10 μm into the sample and it was found quite usual for the sample to lose this amount during a diffusion. The weight loss varied from experiment to experiment and inevitably gave rise to irregular results. The problem was solved by carrying out the diffusions with an excess arsenic pressure of about 0.7 atm in the ampoule. Under these circumstances any weight loss amounted to less than 1 μm and highly reproducible results were obtained. The profiles proved to be insensitive to the amount of tin used in the ampoule over a wide range (1 μg–100 μg in an ampoule of 6 cm^3 volume).

They found that the profiles could be fitted reasonably well by standard erfc solutions, so it was possible to assign values of diffusion coefficient, D, to the curves. Two types of substrate were used: undoped, and n-type with electron concentration

2.4×10^{18} cm^{-3}. An interesting finding was that the value of D was different for the two types of sample. Figure 3.8 shows the plots of D against $1/T$. The results fit two straight lines, line A for the undoped starting material and line B for the n-type. For low diffusion temperatures, D was more than an order of magnitude greater in the doped material than in the undoped. As the diffusion temperature increased, the two values converged so that for temperatures greater than about 1100 °C there was no appreciable difference in the results from the two kinds of sample. It was pointed out by Tuck and Badawi that this convergence appeared to relate to a similar convergence in values of electron concentration, n. At the lower temperatures, the value of n was much greater in the n-type than in the undoped samples. As the temperature was raised, the electron concentration did not change very much in the n-type specimens but increased in the undoped due to thermally generated carriers. At about 1100 °C (Casey 1973) both doped and undoped samples were intrinsic and therefore at temperatures in excess of this there was no difference in the two types of specimen, electrically speaking.

The above argument strongly suggests that D depends primarily on the electron concentration and this led to the proposal that most of the tin exists in the form Sn_{Ga}^{+} and diffuses via gallium vacancies. It seems probable that both gallium and arsenic vacancies in GaAs can be charged and that the gallium vacancy acts as an acceptor, V_{Ga}^{-} (Chiang and Pearson 1975). The equilibrium between the charged and uncharged vacancies can be written

$$V_{Ga}^{0} + e \rightleftharpoons V_{Ga}^{-} \qquad (3.29)$$

where V_{Ga}^{0} is an uncharged vacancy. Equation (3.29) leads to the mass action formulation

$$[V_{Ga}^{-}] = Kn[V_{Ga}^{0}]. \qquad (3.30)$$

To a good approximation $[V_{Ga}^{0}]$ should be constant at constant temperature and fixed ambient vapour pressures, giving $[V_{Ga}^{-}] \propto n$. An increase in electron density therefore gives rise to an increase in diffusion coefficient.

One great attraction of this mechanism, apart from its simplicity, lies in the fact that the ionised donor would be attracted to the charged vacancy by a Coulombic force. Adjacent gallium sites are not, of course, nearest neighbours in the lattice, so any help is likely to be welcomed by the tin atom. There is also some evidence to suggest that the concentration of charged gallium vacancies may be quite large on the arsenic side of the phase diagram (Tuck 1976).

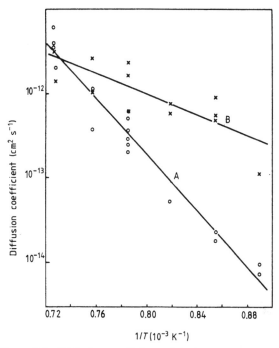

Figure 3.8 Diffusion coefficient of tin in GaAs as a function $1/T$. Curve A is for undoped starting material, curve B for n-type substrates (from Tuck and Badawi 1978).

An alternative model was proposed by Shaw (1984). He took up Panish's (1973) idea of a complex of the form $(V_{Ga}Sn_{As}^-V_{Ga})$ and assumed that this was the major diffusing species. The donor Sn_{Ga}^+ still exists in this model, of course, but as an immobile atom. At first sight, the model may appear to be identical to the one proposed by Young and Pearson (1970) for sulphur in GaAs. There are important differences, however. The atom on the arsenic site is group IV rather than group VI, so is negatively charged. The complex therefore behaves as an acceptor rather than a donor and it follows that the mass action relations governing the formation of the complex are quite different. It is convenient to rename the complex A^-; we can then write

$$[Sn] = [Sn_{Ga}^+] + [A^-]. \tag{3.31}$$

The relationship between the two species can be written

$$Sn_{Ga}^+ + V_{As} + V_{Ga} + 2e \rightleftharpoons A^- \tag{3.32}$$

where e is an electron. Applying mass action to equation (3.32),

$$[A^-] = Kn^2[Sn_{Ga}^+] \tag{3.33}$$

where we have again used the fact that the product of gallium and arsenic vacancies is a constant.

Shaw considered two limiting cases for the model, namely diffusion under gallium-rich and under arsenic-rich conditions. For the former case, he assumed that $[Sn_{Ga}^+] \gg [A^-]$, so the electrical neutrality condition is $n = [Sn_{Ga}^+]$. Equation (3.33) becomes

$$[A^-] = K[Sn_{Ga}^+]^3. \tag{3.34}$$

Using the same argument as that employed for the case of sulphur in GaAs, we can say that the tin diffuses with an effective coefficient given by

$$D = D_A \frac{\partial[A^-]}{\partial[Sn]} \tag{3.35}$$

where D_A is the acceptor diffusion coefficient. With the current set of assumptions, $[Sn] \simeq [Sn_{Ga}^+]$, so using equation (3.34) gives

$$D = 3KD_A[Sn_{Ga}^+]^2 \tag{3.36}$$

i.e. a diffusion coefficient which is strongly dependent on tin concentration. This fits in rather well with some work reported by Arnold *et al* (1984), which was probably carried out under gallium-rich conditions. They used SIMS to determine their diffusion profiles and suggested that the curves corresponded to a diffusion process for which D was proportional to the square of the tin concentration. The amount of data presented in the paper is small, however, and the SIMS profiles cover less than two orders of magnitude in tin concentration.

For arsenic-rich conditions Shaw assumed that complete compensation takes place. This means that $[Sn_{Ga}^+]$ can be put equal to $[A^-]$, and the electron concentration is the intrinsic carrier concentration at the diffusion temperature, n_i. It follows from this assumption that

$$[A] = [Sn_{Ga}] = \tfrac{1}{2} [Sn]$$

and the argument used above leads to an effective diffusion coefficient

$$D = \tfrac{1}{2} D_A \qquad (3.37)$$

which is independent of tin concentration. The mass action constant of equation (3.33) becomes

$$K = 1/n_i^2.$$

This last set of assumptions must be considered slightly suspect. There is little evidence to support the idea that GaAs when heavily doped with tin is likely to be intrinsic at diffusion temperature. In the original paper proposing the existence of the $(V_{Ga}Sn_{As}^-V_{Ga})$ complex, Panish (1973) reported electrical measurements on samples of GaAs grown by liquid phase epitaxy from liquid compositions along the relevant liquidus. His results indicated increasing compensation as the arsenic in the liquid increased. He did not report complete compensation (i.e. intrinsic behaviour) even for the highest arsenic concentrations, however. It is perhaps worth noting here that, strictly, Panish's results for high arsenic composition of the liquid are not relevant to most of the published work on diffusion in GaAs. For the most part, when diffusions are carried out on the arsenic side of the phase diagram, no liquid is produced; the diffusion system consists of solid GaAs and a vapour phase. This point will be considered again in Chapter 4.

Shaw (1984) compared the predictions of equations (3.36) and (3.37) with some results taken from the literature for 950 °C and found agreement. It is difficult, however, to reconcile the results shown in figure 3.8 with equation (3.37). These results were taken under arsenic-rich conditions and show a marked difference between diffusion in undoped and n-type material. According to equation (3.37), no such difference should occur, since both types of sample would have been intrinsic at diffusion temperature. There is good reason to believe, however, that for arsenic-rich experiments the electron concentration can fall below that of the tin. This effect is most marked when the tin concentration is high.

Tuck and Badawi (1978) produced radiotracer and electrical profiles from the same experiments, the latter being determined by a
capacitance–voltage method. An example is shown in figure 3.9.
They found that the electron concentration never exceeded
3.4×10^{18} cm^{-3}, even though the atomic surface concentrations
were normally in the range 10^{19}–10^{20} cm^{-3}. Yamazaki *et al* (1975)
came to a similar conclusion, as did Kressel *et al* (1968) and
Bolkhovityanov and Bolkhovityanova (1975). Arnold *et al*, on the
other hand, worked on the gallium side of the phase diagram and
found no significant difference between the two concentrations.
The tin concentrations in their experiments, however, were rather
lower than those reported in the other studies.

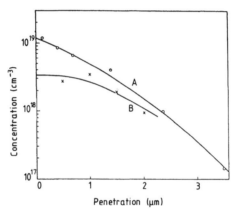

Figure 3.9   Tin concentration (curve A) and electron concentration (curve B) for two specimens diffused in the same
ampoule (from Tuck and Badawi 1978).

Tin diffusion has also been studied in InSb by Sze and Wei (1961)
and in GaSb by Uskov (1974). In the latter piece of work,
radioactive tin was diffused into n-type material doped with
tellurium, and also into p-type doped with zinc. Experiments were
carried out in the temperature range 500 °C–650 °C and profiles
were plotted. The surface concentrations of the profiles were in the
region of 10^{21} cm^{-3}, which is surprisingly high. The profile shapes
did not correspond to any of the well-known solutions of Fick's
law. At the higher temperatures used, the diffusion in n-type
material was slower than in the p-type; this was attributed to the

formation of SnTe complexes, although no direct evidence was produced for the existence of this type of defect.

Sze and Wei (1961) also used radiotracer tin for their InSb study. Experiments were carried out between $385\,^{\circ}C$ and $512\,^{\circ}C$. In all cases a value of D could be allocated to the profile and a plot of D v. $1/T$ gave a straight line of the form of equation (2.2), with $D_0 = 5.5 \times 10^{-8}\,cm^2\,s^{-1}$ and $Q = 0.75\,eV$. In the same set of experiments they also diffused radioactive zinc into InSb. The surprising result was obtained that over the whole range studied, the tin diffusion coefficient was about twice that shown by zinc.

3.4 Silicon and Germanium

High temperature diffusion of silicon in GaAs was first studied by Vieland (1961) and Antell (1965), both using the p–n junction technique to assess depth of diffusion. Vapour and solid sources, respectively, were used by these workers and both sets of data indicated that increasing the ambient arsenic pressure brought about an increase in diffusion penetration. A rather similar result was reported by Kasahara and Watanabe (1982). They implanted silicon into GaAs and annealed at $900\,^{\circ}C$. Carrier concentration profiles were plotted and diffusion coefficients were estimated from the profile shapes. Experiments were performed using a variety of values of arsine partial pressure. They found that a rise of arsine pressure corresponded to an increase in estimated diffusion co-efficient. This set of results must be interpreted with some care. Diffusion which occurs during post-implantation annealing is not 'normal' diffusion, since it usually takes place in damaged crystal. The use of carrier concentration profiles rather than atomic profiles is also dangerous here; the two could be quite different, since silicon can, in principle, occupy either lattice site and therefore give rise to acceptors as well as donors. Nevertheless, it is interesting that the qualitative result reported by Kasahara and Watanabe agrees with that of the earlier workers.

A simple model for the diffusion of amphoteric impurities was put forward by Greiner and Gibbons (1984, 1985) and was used to account for the diffusion of silicon in GaAs. It was proposed that three species co-exist in the GaAs lattice: a silicon atom on a gallium site, acting as a donor, Si_{Ga}^{+}, a silicon atom on an arsenic site, acting as an acceptor, Si_{As}^-, and the neutral complex $(Si_{Ga}Si_{As})$. It was pointed out by these authors that, although silicon atoms can exist on either site, any diffusion mechanism involving motion of individual atoms through both gallium and arsenic vacancies would require changes in charge state as well as vacancy type. This would inevitably lead to a very low diffusion coefficient. Progression of the neutral complex is relatively simple, however, since no charge exchange is involved. If a gallium vacancy moves to a site adjacent to the complex, movement can take place as follows:

$$(Si_{Ga}Si_{As}) + V_{Ga} \rightleftharpoons V_{Ga} + (Si_{As}Si_{Ga}). \qquad (3.38)$$

The next move requires the assistance of an arsenic vacancy:

$$(Si_{As}Si_{Ga}) + V_{As} \rightleftharpoons V_{As} + (Si_{Ga}Si_{As}) \qquad (3.39)$$

and so on.

A number of very short diffusions of silicon into GaAs (10–300 s) was carried out by Greiner and Gibbons over the temperature range 850–1050 °C. They used SIMS to determine the atomic profile and electrical techniques for the electron profile. A result is shown in figure 3.10. It can be seen that the silicon gives n-type doping, but that the concentration of donors is generally less than the atomic concentration, as would be predicted by their theory. They gave the following analysis of the model.

The equation relating the complex to the charged species is

$$Si_{Ga}^+ + Si_{As}^- \rightleftharpoons (Si_{Ga}Si_{As}) \qquad (3.40)$$

giving rise to the mass action relation

$$[Si_{Ga}^+] [Si_{As}^-] = K[C] \qquad (3.41)$$

where the complex is re-named C to make the nomenclature simpler. The total amount of silicon must be equal to the sum of that contained in the three species, i.e.

$$[Si] = [Si_{Ga}^+] + [Si_{As}^-] + 2[C]. \qquad (3.42)$$

Greiner and Gibbons then make a rather large assumption. Because of the low electron concentration indicated in figure 3.10, they put

$$[Si_{Ga}^+] = [Si_{As}^-]. \qquad (3.43)$$

This is not the only condition which would give a significant difference between the atom and electron concentrations; the condition

$$[C] \gg [Si_{Ga}^+], \gg [Si_{As}^-]$$

for instance, would have the same result. Equation (3.43) can, in any case, never be completely true; if it were, the sample would become intrinsic rather than n-type. At lower concentrations it is certainly incorrect, since the discrepancy between the atomic and carrier concentrations disappears. However, we continue the argument, using equation (3.43), remembering that it is acceptable as an approximation only at the highest concentrations.

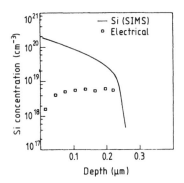

Figure 3.10 SIMS and electrical profiles for silicon diffused into GaAs at 1050 °C for 3 s (from Greiner and Gibbons 1984).

Equations (3.41)–(3.43) combine to give the concentration of the complex in terms of the total silicon concentration:

$$\frac{2[C]}{[Si]} = 1 + \frac{K}{[Si]} \left\{ 1 - \left(1 + \frac{2[Si]}{K}\right)^{1/2} \right\}. \qquad (3.44)$$

The fraction of silicon tied up in the complex depends very much on the value chosen for K, of course; for $K = [Si]$, for instance, it amounts to 27 percent. Assuming that dopant pairs dominate the diffusion process, we can use the argument which leads to equation (3.13) to obtain an expression for the effective diffusion coefficient of silicon:

$$D = 2D_c \frac{\partial[C]}{\partial[Si]} \qquad (3.45)$$

where D_c is the diffusion coefficient of the complex and the right-hand side is multiplied by two because the flux of silicon atoms is twice the flux of the complex. Combining equations (3.44) and (3.45) gives the final expression for the effective diffusion coefficient:

$$D = D_c \left(1 - \frac{1}{(1 + 2[Si]/K)^{1/2}}\right). \qquad (3.46)$$

This is concentration-dependent, tending to a constant value for [Si] $\gg$ K. At this stage all the silicon would be in the form of the complex, and the dopant would have no electrical effect on the semiconductor. Greiner and Gibbons (1984) substituted equation (3.46) into Fick's law and solved the resulting differential equation numerically. Figure 3.11 shows the SIMS result of figure 3.10 together with a calculated profile obtained for $K = 4 \times 10^{19}$ cm^{-3} and $D_c = 5 \times 10^{-11}$ cm^2 s^{-1}. The calculated curve fits the results well for a concentration range covering three orders of magnitude. It should be noted, however, that at the most generous estimate, the approximations of the theory render the model valid only for concentrations above about 2×10^{19} cm^{-3}.

At first sight, the theory does not predict the experimental result that increasing the arsenic overpressure increases the silicon diffusion rate. Greiner and Gibbons (1985) put forward the following argument by way of explanation. According to the model, silicon pairs move through the crystal by alternating exchanges with gallium and arsenic vacancies. The rate at which these exchanges occur will depend on the number of silicon pairs which have the energy required for the exchange and also on the fraction of time that an adjacent lattice site is vacant. The first of these effects is dependent on temperature and the second is proportional to the gallium and arsenic vacancy concentrations. It is believed that the concentration of arsenic vacancies in GaAs exceeds that of gallium vacancies (Potts and Pearson 1966, Logan and Hurle 1979). It would seem reasonable to assume, therefore, that the waiting time for a gallium vacancy would be longer than that for an arsenic vacancy. The rate-limiting step in the diffusion process is therefore likely to be equation (3.38) rather than equation (3.39). Any change in the experimental conditions causing [V_{Ga}] to increase might be expected to create a greater diffusion rate. Making the arsenic overpressure larger gives rise to such an increase in vacancies. This argument essentially makes the quantity D_c in equation (3.46) a function of [V_{Ga}] and hence of arsenic vapour pressure, rather than a simple constant.

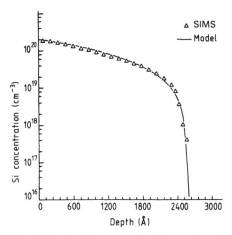

Figure 3.11   Calculated curve using the model of Greiner and Gibbons, fitted to the results of figure 3.10 (from Greiner and Gibbons 1985).

The above makes an intriguing argument. Unfortunately, even less is known about values of the energy barriers involved than is known about values of the precise concentrations of vacancies in GaAs. What we do know is that, given the choice, single silicon atoms prefer to sit on gallium sites when silicon is diffused into GaAs, i.e. [Si_{Ga}^+] > [Si_{As}^-]; if this were not so, silicon-doped material would be p-type rather than n-type. This happens despite the presumed excess of arsenic vacancies and could indicate that the energy barrier for the reaction of equation (3.38) is less than that for equation (3.39).

The Greiner–Gibbons theory is so simple and reasonable that it is tempting to consider it as a general theory for the diffusion of amphoteric impurities in the III–Vs and, in particular, for the tin data described earlier. Unfortunately, it would not explain the result shown in figure 3.8; there is nothing in the theory described above which would predict a diffusion coefficient dependent on the level of doping.

It has been pointed out that a given group IV atom will have a preference for sitting on the group III site or the group V site. There are probably many reasons for this, but one of them is likely to be the size of the atom compared to the size of the one it replaces. The covalent radii of gallium and arsenic are, respectively, 1.26 Å and 1.18 Å (Madelung 1964). Tin is a relatively large atom, of covalent radius 1.40 Å, so it is perhaps not surprising that it tends to replace the larger of the two host atoms, giving rise to the donor Sn_{Ga}^+. This consideration would not stop the proposed acceptor ($V_{Ga}Sn_{As}^-V_{Ga}$) forming, of course; with a vacant site on each side of it, the tin atom would be quite comfortable. Silicon, on the other hand, at 1.17 Å, is closer in size to the host atoms, so one might expect to find it occupying both sites, as suggested by the Greiner–Gibbons model.

Diffusion of germanium into GaP was studied by Schneider and Nebauer (1975) at 900 °C and 1000 °C. Radioactive germanium was used, diffusion profiles were plotted and 'best-fit' erfc curves were drawn through the experimental points. The points nearest the surface did not lie on the curves, however, so the values of D produced by this work must be regarded as approximations. Experiments were carried out over a range of ambient phosphorus pressure and it was found that the diffusion coefficient varied inversely as the square of the phosphorus pressure. This result led them to the conclusion that diffusion was taking place via phosphorus vacancies. A model was proposed in which the germanium exists on both gallium and phosphorus sites as, respectively, donors and acceptors, with $[Ge_{Ga}^+] \simeq [Ge_P^-]$. The germanium on gallium sites was assumed to be immobile, although no reason was given why this should be so. No electrical measurements were reported in this work and this is unfortunate, since it is likely that the electrical state of germanium in GaP is concentration dependent. Trumbore *et al* (1965) reported that doping with germanium yields p-type material at the lower doping levels and n-type at higher levels, with nearly equal numbers of germanium atoms on the two sites at all levels of doping. This behaviour would almost certainly have an interesting effect on the diffusion mechanism.

Chapter 4

Shallow Acceptors, especially Zinc

Atoms from group II of the Periodic Table normally occupy group III sites in III–V semiconductors, acting as shallow acceptors. Zinc is the most commonly used acceptor, and Zn/GaAs is one of the best documented of all semiconductor systems. Group II atoms diffuse rather faster than the atoms considered in Chapter 3, largely because of the role played by interstitials. In many cases, the 'substitutional–interstitial' diffusion mechanism, or one of its variations, has been found to operate. Because of its importance, we start by giving the basic theory for this mechanism before considering specific cases. Deviations from the basic model will be considered as they arise.

4.1 The Substitutional–Interstitial Mechanism

This model was originally proposed by Frank and Turnbull (1956) to explain the anomalous diffusion of copper in germanium. It was suggested by Longini (1962) that the same mechanism might account for zinc diffusion in GaAs. Consider a group II impurity, most of which exists substitutionally on the group III site, acting as a singly ionised acceptor. Call this substitutional acceptor s, and let its concentration be S (it is necessary, for reasons which will become obvious, to adopt a very simple nomenclature at this point). The model assumes that a small proportion of the element exists interstitially; assume that it acts as a singly ionised donor, i, of concentration I. The substitutional form has a negligible diffusion coefficient and diffusion proceeds, by movement of the highly mobile interstitial atom. The process turns out to be concentration-dependent because the proportion of the dopant which is in the mobile form increases with total dopant concentration. A simple relationship exists describing the transition between the two forms:

$$s^- + 2h \rightleftharpoons i^+ + v_{III} \qquad (4.1)$$

where v_{III} is a vacancy on the group III site and h is a hole. If we can make the assumption that this reaction comes to equilibrium at a rate which is fast compared to any diffusion rate that is operating, the law of mass action can be applied:

$$K_1 p^2 S = IV \qquad (4.2)$$

where V is the group III vacancy concentration. Equation (4.2) is true throughout the crystal. By postulate, $S \gg I$ so, in order for the crystal to remain neutral, $p \simeq S$ and equation (4.2) becomes

$$K_1 S^3 = IV. \qquad (4.3)$$

Now consider a diffusion process in which dopant is introduced into a semiconductor slice from an external phase, maintaining the surface concentrations constant. If these surface concentrations are indicated by a prime, then at the surface we have the special case of equation (4.3):

$$K_1 S'^3 = I' V'. \qquad (4.4)$$

We now make the assumption that in the bulk of the crystal the vacancy concentration is maintained at the surface value, V', throughout the diffusion process. This is a fairly gross assumption which will be reconsidered in a later section. The substitutional and interstitial concentrations, on the other hand, are functions of x and t so that inside the crystal equation (4.3) becomes

$$K_1 S^3(x, t) = I(x, t) V' \qquad (4.5)$$

where $x = 0$ corresponds to the surface. Since the substitutional species is immobile, any increase in either substitutional or interstitial dopant inside the semiconductor is due to diffusion of interstitial atoms, i.e.

$$\frac{\partial I}{\partial t} + \frac{\partial S}{\partial t} = D_i \frac{\partial^2 I}{\partial x^2} \qquad (4.6)$$

where D_i is the diffusion coefficient of interstitials. The quantity I can be replaced by S using equation (4.5), giving

$$\left(\frac{3K_1 S^2}{V'} + 1\right)\frac{\partial S}{\partial t} = D_i \frac{\partial}{\partial x}\left(\frac{3K_1 S^2}{V'}\frac{\partial S}{\partial x}\right). \qquad (4.7)$$

If we substitute for K on the left-hand side using equation (4.4), it is seen that the first term in the bracket is small compared to unity. Equation (4.7) therefore simplifies to

$$\frac{\partial S}{\partial t} = \frac{\partial}{\partial x}\left(\frac{3K_1 D_i}{V'}S^2\frac{\partial S}{\partial x}\right). \qquad (4.8)$$

This is of the form of Fick's law, with

$$D = \frac{3K_1 D_i S^2}{V'} \qquad (4.9)$$

i.e. a diffusion coefficient is obtained which is proportional to the square of the substitutional concentration. Since, by postulate, nearly all of the dopant is substitutional, we effectively have $D \propto C^2$, where C is the total dopant concentration. A careful reading of the above will show the reader that the power of two which appears in equation (4.9) corresponds to the difference of two in the charge states of the substitutional and interstitial atoms. Thus if we had chosen the initial donor to be doubly charged we would have found $D \propto C^3$.

It should be noted that in the above derivation it has been assumed that the crystal is extrinsic and that the carrier concentrations are therefore determined by the acceptor doping. If the intrinsic carrier concentration, n_i, is substantially greater than S at diffusion temperature, the crystal is intrinsic and the substitution for p in equation (4.2) becomes $p = n_i$. Following through the above argument for this case gives a value for diffusion coefficient which is independent of acceptor concentration (Tuck 1969). This is by no means an insignificant consideration, since n_i can become quite large at the elevated temperatures normally used for diffusion; in GaAs at 1000 °C, for example, it is about 2×10^{17} cm^{-3}.

4.2   Zinc in GaAs

4.2.1   Diffusion results and models

Early work in which diffusion profiles were measured (Cunnel and Gooch 1960, Goldstein 1961) indicated that they did not correspond to any of the well-known Fick's law solutions. The rate of diffusion appeared to depend on concentration and profiles often showed concave sections. The diffusion front was invariably abrupt. Typical examples are shown in figure 4.1. One obvious approach is to use the Boltzmann–Matano method, described in Chapter 2, to determine the form of the function $D(C)$ from the experimental profiles. This was tried by a number of workers with rather mixed and unreproducible results. Eventually, Tuck and Kadhim (1972) carried out a series of experiments in which radiotracer zinc profiles were plotted for identical diffusion conditions but varying diffusion times. The work showed that the profiles were not proceeding into the crystal as $x/t^{1/2}$ and it therefore followed that the Boltzmann–Matano method of analysis was not valid for this system.

If the Boltzmann–Matano technique lets us down, there is little alternative but to use the rather more time-consuming isoconcentration method. Isoconcentration work was carried out by Chang and Pearson (1964b) but the quantity of data produced was rather too small to give an unambiguous indication of the form of $D(C)$. Later work by Kadhim and Tuck (1972) increased the available data; their result shown in figure 4.2 shows $D \propto C^n$, with n close to a value of 2 for a diffusion temperature of $1000\,°C$. It would appear, therefore, that the theory as outlined in section 4.1, with singly-charged substitutional and interstitial atoms, applies in this case. It follows that equation (4.9) is appropriate with S, the concentration of zinc acceptors, essentially equal to the total zinc concentration, and V' equal to the equilibrium concentration of gallium vacancies.

Although we would hope that in a given isoconcentration experiment the quantity $[V_{Ga}]$ would remain constant, there is some scope for varying it by changing the experimental conditions. This can only be done within certain limits; the extent to which GaAs can vary from a strict 1 : 1 concentration ratio of gallium to arsenic atoms is fixed by the phase diagram. Within these limits,

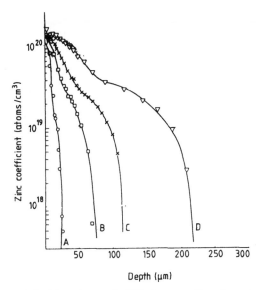

Figure 4.1 Experimental diffusion profiles for zinc in GaAs at 1000 °C with excess arsenic in ampoule. Diffusion times: A, 10 min; B, 90 min; C, 3 h; D, 9 h (from Tuck and Kadhim 1972).

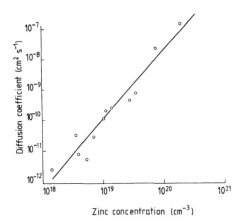

Figure 4.2 Variation of diffusion coefficient of zinc in GaAs as a function of zinc concentration. The results were taken at 1000 °C at dissociation pressure of arsenic (from Kadhim and Tuck 1972).

however, [V_{Ga}] can be altered by varying the ambient arsenic pressure. This was considered in deriving equation (3.3), which is repeated here for convenience:

$$[V_{Ga}] = V' = K_2 P_{As_4}^{1/4}. \tag{4.10}$$

Comparing equations (4.9) and (4.10) we see that increasing the ambient arsenic vapour in a zinc diffusion experiment should bring about a reduction in measured diffusion coefficient. This prediction was checked by Kadhim and Tuck (1972) who repeated the experiments of figure 4.2 using, this time, a high arsenic vapour pressure inside the diffusion ampoule. Their results are shown in figure 4.3. Comparison of the two figures shows that the expected reduction is indeed found; Kadhim and Tuck demonstrated that the order of magnitude of the reduction is consistent with the above theory.

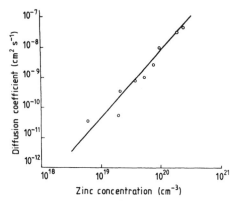

Figure 4.3   Variation of diffusion coefficient of zinc in GaAs as a function of zinc concentration. The results were taken at 1000 °C at an excess arsenic pressure of 4×10^{-2} atm (from Kadhim and Tuck 1972).

Having used the isoconcentration technique to obtain D as a function of C it should be a simple matter to substitute $D(C)$ into Fick's law and obtain the form of profiles determined in non-isoconcentration (or 'chemical') diffusions. This calculation was made by Weisberg and Blanc (1963) and turned out to be rather disappointing. The calculated profiles do not give a good fit to experimental curves such as those of figure 4.1; perhaps this is not too surprising in view of the fact that their curves scale as $x/t^{1/2}$, while, as noted above, experimental profiles do not. It seems likely that some assumption which is valid in the case of isoconcentration diffusions is invalid for chemical diffusions. The basic difference between the two types of experiment is that isoconcentration diffusions take place at thermodynamic equilibrium, while chemical diffusions do not. The possibility arises, therefore, of defect equilibrium breaking down in the latter case so that, for instance, equation (4.5) may not apply for chemical diffusions.

It was pointed out by Tuck (1969) and by Shaw and Showan (1969a) that every time equation (4.1) operates in the bulk of the semiconductor, the crystal loses a gallium vacancy. The crystal will try to maintain its equilibrium vacancy concentration by some mechanism such as dislocation climb, but if it is not completely successful in this, V will fall below V' inside the semiconductor. This idea was pursued by Tuck (1971) and by Tuck and Kadhim (1972). They assumed that the crystal had some mechanism in the bulk which produced vacancies at a rate proportional to the shortfall. Thus, in addition to equations (4.2) and (4.6) a further differential equation must be employed to describe the concentration of vacancies at distance x from the surface

$$\frac{\partial V}{\partial t} = D_v \frac{\partial^2 V}{\partial x^2} - \frac{\partial S}{\partial t} + k(V' - V) \qquad (4.11)$$

where k is a constant and D_v is the diffusion coefficient for vacancies. The first term on the right-hand side represents vacancies gained from the surface, the second represents loss of vacancies due to interstitial zinc going substitutional and the third is the bulk production of vacancies by dislocation climb etc. A complete solution of the model therefore requires the simultaneous solution of equations (4.2), (4.6) and (4.11) together with a fourth which requires that at any point the concentrations of gallium atoms, gallium vacancies and substitutional zinc atoms add up to the concentration of group III sites in the crystal. By making a (fairly large) number of simplifying assumptions, Tuck and Kadhim were able to show that the model predicts profiles of the correct form.

Recently an interesting variation on the above approach has been proposed by Gosele and Morehead (1981). Their mechanism, which they describe as a 'kick-out' model, has the diffusing interstitial zinc atom joining the lattice by pushing a gallium atom off its site and creating a gallium interstitial, i.e. equation (4.1) is replaced by

$$Zn_i^+ \rightleftharpoons Zn_{Ga}^- + I_{Ga} + 2h. \qquad (4.12)$$

At thermodynamic equilibrium (e.g. during an isoconcentration diffusion) the two approaches are indistinguishable, since the gallium interstitials and vacancies are related by

$$Ga_{Ga} \rightleftharpoons I_{Ga} + V_{Ga} \qquad (4.13)$$

and equation (4.9) is still obtained for diffusion coefficient. In the vacancy model, the crystal has the problem of producing vacancies; in the kick-out model the problem is to eliminate excess gallium interstitials. In either case, dislocations presumably have a role to play. Gosele and Morehead gave a semi-quantitative analysis which indicated that profiles similar to those obtained experimentally are predicted by the model.

The two models were compared in detail by van Ommen (1983) who used both approaches to fit theoretical profiles to experimental results of Tuck and Kadhim (1972) and Ting and Pearson (1971). He made similar assumptions about the elimination of excess gallium interstitials to those previously made about production of vacancies, i.e. he assumed the rate of elimination was proportional to the excess of interstitials over the thermal equilibrium value. He also found that both models gave profiles of the right shape but concluded that the kick-out model gave the better fit. It should be noted, however, that results obtained from both of these models depend critically on the assumptions made about the rate of elimination of interstitials on the one hand and production of vacancies on the other. These assumptions have been somewhat arbitrary in all of the modelling carried out to date and the theoretical results can be considered only semi-quantitative.

The best evidence presented so far in support of a kick-out mechanism has come from metallurgical examination of zinc-diffused samples. It was established early on that the diffusion of zinc into GaAs could give rise to precipitation (Black and Jungbluth 1967a) and dislocations (Schwuttke and Rupprecht 1966, Black and Jungbluth 1967b). In more recent work, Ball *et al* (1981) and Hutchinson and Ball (1982) carried out careful transmission electron microscope studies on zinc-diffused GaAs. They found faulted and perfect dislocation loops in the diffused region and identified them as being interstitial in nature. The defects were at their highest concentration near to the diffusion front. Ball *et al* suggested a mechanism which is essentially a compromise between the two available versions of the model. It was proposed that the diffusing zinc joins the lattice according to the substitutional–interstitial relation of equation (4.1). This causes a shortfall of vacancies which is relieved by gallium vacancies going interstitial, i.e. by the operation of equation (4.13). This way of looking at the process makes equation (4.13) the vacancy generation mechanism mentioned, but not specified, by Tuck and Kadhim (1972).

The approximate consensus on the diffusion mechanism of zinc in GaAs has recently been questioned by Shaw (1984). He shows that an alternative mechanism, based on the donor complex $(Zn_{Ga}V_{As})^+$ would give similar results to both of the available substitutional–interstitial models. Again it is assumed that most of the zinc acts as an acceptor on the gallium site and that it is immobile in that state. The complex moves through the crystal according to the sequence

$$Zn_{Ga}V_{As} \rightarrow V_{Ga}Zn_{As} \rightarrow Zn_{As}V_{Ga} \rightarrow V_{As}Zn_{Ga} \rightarrow Zn_{Ga}V_{As}. \quad (4.14)$$

$$(1)\,(2) \qquad (1)\quad(2) \qquad (2)\,(3) \qquad (2)\,(3) \qquad (3)\,(4)$$

The numbers in brackets label the lattice sites occupied by the atoms and vacancies. The two species are related by the equation

$$Zn_{Ga}^- + V_{As} \rightleftharpoons (Zn_{Ga}V_{As})^+ + 2e. \quad (4.15)$$

Applying the law of mass action

$$[Zn_{Ga}^-][V_{As}] = K_i n^2 C^+. \quad (4.16)$$

where C^+ represents the complex. We also have

$$np = \text{constant} \quad (4.17a)$$

and

$$[V_{Ga}][V_{As}] = \text{constant} \quad (4.17b)$$

and the fact that charge neutrality of the semiconductor requires

$$[Zn_{Ga}^-] = p. \quad (4.18)$$

Substituting equations (4.17) and (4.18) into (4.16) gives

$$[V_{Ga}]C^+ = K_2[Zn_{Ga}^+]^3. \quad (4.19)$$

This is of the same form as equation (4.4) and leads to an expression for diffusion coefficient similar to equation (4.9). On the whole the mechanism seems rather unsatisfactory as an explanation when compared to the interstitial approach and firm experimental evidence will be required before it replaces the earlier models.

4.2.2   Computer modelling of mechanism

A complete analysis of the substitutional–interstitial diffusion mechanism for zinc in GaAs involves the simultaneous solution of four equations, two of which are partial differential equations. This is rather a tall order, and none of the attempts at modelling have quite achieved it. Both Gosele and Morehead (1981) and Tuck and Kadhim (1972), for instance, divided the crystal into regions in which, to a reasonable approximation, rather simpler versions of their models applied. This approach to analysing the substitutional–interstitial mechanism will be pursued in the following chapter when the diffusion of chromium in GaAs is considered. The closest approach to a complete solution for the zinc/GaAs system, however, was carried out by Zahari and Tuck (1985), using the direct modelling approach which has been outlined in section 2.3 and was used in section 3.1 for the sulphur/GaAs system. The only simplification made by these workers was to ignore the charge on the substitutional atoms, so that equations (4.1) and (4.2) became

$$i + v_{III} \rightleftharpoons s \tag{4.20}$$

$$K_2 IV = S \tag{4.21}$$

where, again, it is understood that the concentrations I, V, S are all functions of distance and time. At the surface,

$$K_2 I' V' = S'. \tag{4.22}$$

Crystal 'planes', or nodes, are set up and numbered $1 \ldots, (n-1), n, (n+1), \ldots$ as in figure 2.5. Each node represents a region of crystal containing L gallium sites. The node at $n = 1$ corresponds to the semiconductor surface region. Each lattice point is occupied by a host atom, a substitutional impurity atom or a vacancy, so for all n,

$$L = H_n + S_n + V_n \tag{4.23}$$

where H_n, S_n, V_n are the concentrations of host atoms, substitutional impurities and vacancies respectively at the nth node. The interstitial impurity atoms do not occupy lattice points and thus are not included in the above equation.

As before, it is assumed that the substitutional atom is immobile so only movement of the host atoms (or, to put it another way, of vacancies) and of interstitial atoms need be considered. Diffusion of the latter is simple since the concentration of empty interstitial sites is large compared with the concentration of interstitials. A relation of the form of equation (2.42) is followed:

$$I_n^+ = I_n + k_i(I_{(n+1)} - 2I_n + I_{(n-1)}) \qquad (4.24)$$

where

$$k_i = \frac{D_i \Delta t}{(\Delta x)^2} \qquad (4.25)$$

and D_i is the diffusion coefficient of the impurity atoms. Self-diffusion is less simple and due allowance must be made for the fact that a host atom can only jump if there is a vacant site on the lattice for it to occupy. The same situation was considered in the S/GaAs case, and equation (3.20) is appropriate:

$$H_n^+ = H_n + k_h(H_{(n-1)}V_n + H_{(n+1)}V_n - H_n V_{(n-1)} - H_n V_{(n+1)}) \qquad (4.26)$$

where

$$k_h = \frac{D_h \Delta t}{V'(\Delta x)^2} \qquad (4.27)$$

and D_h is the self-diffusion coefficient for the host atoms (gallium atoms in this case).

The production of extra vacancies is performed using the Tuck and Kadhim (1972) hypothesis that they are created at the node n at a rate which is proportional to the short-fall. In time Δt, ΔV_n vacancies are produced:

$$\Delta V_n = k_v \Delta t (V' - V_n). \qquad (4.28)$$

Zahari and Tuck interpreted this physical process as host atoms moving to interstitial positions.

The computational procedure is as follows. At time $t = 0$, the nodes are set up with

$$S_1 = S' \quad V_1 = V' \quad I_1 = I' \tag{4.29}$$
$$S_n = 0 \quad V_n = V' \quad I_n = 0 \quad n > 1. \tag{4.30}$$

The values of H_n at each node are determined by using equation (4.23). The interstitial and host atoms are allowed to diffuse for the iteration time Δt (note that these two processes are independent of each other) and new concentration values are found using equations (4.24) and (4.26). Because of the imposed boundary conditions the values at node 1 are unchanged. For all other nodes, however, the new concentrations no longer obey equation (4.21). Equation (4.20) is therefore brought into operation, transferring just that number of atoms necessary to make equation (4.21) balance. Finally ΔV_n vacancies are added to the nodes. The sequence of interstitial diffusion, self diffusion, atom transfer and vacancy generation is repeated for each successive time step Δt, taking care that equation (4.23) is never violated.

The model was run using values appropriate to the Zn/GaAs system; typical computed profiles are shown in figure 4.4. The curves compare well with the experimental profiles of figure 4.1.

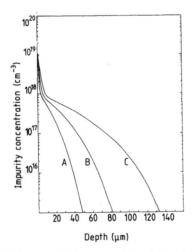

Figure 4.4 Computed zinc impurity profiles for different diffusion times: A, 10 min; B, 30 min; C, 90 min (from Zahari and Tuck 1985).

The main difference is that the experimental profiles show a sharper fall for low concentrations of zinc. This is to be expected in view of the fact that the charges on the substitutional and interstitial atoms were ignored in this treatment.

4.2.3 *The Zn/GaAs phase diagram*

For any diffusion system, a knowledge of the relevant phase diagram can help in two quite different types of situation. First, if we are carrying out experiments attempting to identify the diffusion mechanism involved, we require information on the amount and composition of all phases within the diffusion ampoule; some (but not all) of this information is given by the phase diagram. Second, if we are performing a large number of diffusions, we require a result which is not affected by minor variations in the composition of the diffusion source. Here the diagram can be of great assistance.

For high-temperature diffusion of zinc in GaAs, the information is available, in principle, for establishing the composition and quantity of all phases present in the diffusion enclosure. The phase diagram gives data on the condensed phases (Panish 1966a, 1966c) and data on the various vapour pressures has been gathered by Shih *et al* (1968a). Consider an experiment in which a slice of GaAs is contained in a sealed ampoule, volume V, together with a diffusion source which can, in general, consist of known amounts of gallium, arsenic and zinc. The ampoule is then raised to a high temperature so that diffusion can proceed; 1050 °C will be chosen here, since the phase diagram, shown in figure 4.5, is fairly simple at this temperature. The other data required are the variation of arsenic vapour with liquidus composition, shown in figure 4.6, and the zinc vapour pressure v. composition curve of figure 4.7. We start by calculating the pressure of As_4 in the ampoule; this is given by the ideal gas law:

$$P_{As_4} V = nRT \qquad (4.31)$$

where R is the gas constant, T is the absolute temperature and n is the number of moles of As_4 vapour in the ampoule. The problem is therefore to determine n in terms of the amounts of material in the ampoule.

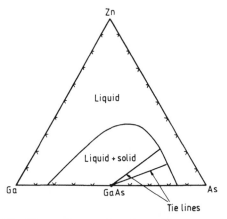

Figure 4.5 Phase diagram for the gallium–arsenic–zinc system at 1050 °C (from Panish 1966a). Reproduced by permission of *Pergamon Journals Ltd*.

Let the weights of added zinc, gallium and arsenic be W_{Zn}^a, W_{Ga}^a and W_{As}^a respectively. At diffusion temperature some of the GaAs disproportionates. GaAs is always close to stoichiometry, so to a good approximation the number of moles of gallium leaving the solid, S say, is equal to the number of moles of arsenic leaving. The

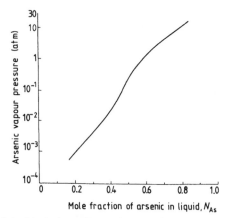

Figure 4.6 Variation of arsenic vapour pressure with the composition of the liquid with which it is in equilibrium. Graph plotted at 1050 °C (from Shih *et al* 1968a). Reproduced by permission of *Pergamon Journals Ltd*.

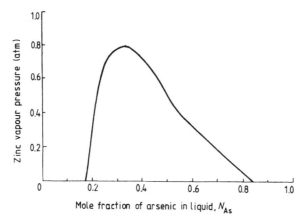

Figure 4.7 Variation of zinc vapour pressure with the composition of the liquid with which it is in equilibrium. Graph plotted for 1050 °C (from Shih *et al* 1968a). Reproduced by permission of *Pergamon Journals Ltd*.

vapour pressure of gallium is known to be negligibly small, so it can be assumed that all of the gallium leaving the GaAs goes into the liquid phase, together with W_{Ga}^a. If there are T moles in the liquid and the mole fraction of gallium is N_{Ga}, it follows that

$$\frac{W_{Ga}^a}{M_{Ga}} + S = N_{Ga} T \qquad (4.32)$$

where M_{Ga} is the atomic weight of gallium.

Of the arsenic that leaves the GaAs, some goes into the liquid and some into the vapour phase. The amount of arsenic in these two phases is supplemented by the W_{As}^a which has been added, amounting to an extra W_{As}^a/M_{As} moles, where M_{As} is the atomic weight. The total amount of zinc in both phases combined is therefore

$$\frac{W_{As}^a}{M_{As}} + S \text{ moles.} \qquad (4.33)$$

Of this, the amount in the liquid phase is $N_{As} T$ moles, where N_{As} is the mole fraction of arsenic in the liquid. So the number of moles of arsenic in the vapour must be

$$\frac{W_{As}^a}{M_{As}} + S - N_{As} T. \qquad (4.34)$$

Substituting for S from equation (4.32) gives for the arsenic in the vapour phase

$$\frac{W_{As}^a}{M_{As}} + T(N_{Ga} - N_{As}) - \frac{W_{Ga}}{M_{Ga}} \text{ moles.} \qquad (4.35)$$

We know that the predominant vapour species at $1050\,^{\circ}\mathrm{C}$ is not As, but As_4. It takes four moles of the former to make one of the latter, so the number of As_4 moles in the vapour is given by

$$n = \frac{1}{4}\left[\frac{W_{As}^a}{M_{As}} + T(N_{Ga} - N_{As}) - \frac{W_{Ga}}{M_{Ga}}\right]. \qquad (4.36)$$

It remains to obtain an expression for T. This can be done in terms of the known added weight of zinc, W_{Zn}^a. Because of the low solubility of zinc in GaAs shown in the phase diagram, we can ignore the zinc in the solid phase. The weight of zinc in the liquid is given by $(W_{Zn}^a - W_{Zn}^v)$ where W_{Zn}^v is the weight in the vapour. The number of moles of zinc in the liquid is found by dividing this weight by the atomic weight of zinc, M_{Zn}. The mole fraction of zinc in the liquid can then be written

$$N_{Zn} = \frac{W_{Zn}^a - W_{Zn}^v}{M_{Zn}T} \qquad (4.37)$$

or,

$$T = \frac{W_{Zn}^a - W_{Zn}^v}{N_{Zn}M_{Zn}}. \qquad (4.38)$$

Substituting equations (4.36) and (4.38) into equation (4.31),

$$P_{As_4} = \frac{RT}{4V}\left[\frac{W_{As}^a}{M_{As}} + \frac{W_{Zn}^a - W_{Zn}^v}{N_{Zn}M_{Zn}}(N_{Ga} - N_{As}) - \frac{W_{Ga}}{M_{Ga}}\right] \qquad (4.39)$$

which is the required result.

The arsenic and zinc vapour pressures and the composition of the liquid can be found by an iterative procedure (Shih *et al* 1968b), using equation (4.39). First a guess is made as to where in figure 4.5 the system lies. Values of N_{Ga}, N_{As}, N_{Zn} are read from the liquidus. Figure 4.7 gives the zinc vapour pressure for this liquid composition, and this information, together with the known volume of the ampoule, allows W_{Zn}^v, the weight of zinc in the vapour, to be calculated. All parameters in equation (4.39) now have values and an external partial pressure of arsenic in the ampoule can be calculated.

A second value for P_{As_4} can be obtained from figure 4.6, using N_{As}. The two estimates for P_{As_4} coincide only if the original guess was correct. A more informed guess is then made and the procedure is repeated. Usually about three cycles gives a consistent result, and the vapour pressures of arsenic and zinc and the liquidus composition are known for the experimental conditions.

As the temperature is reduced, the Ga–As–Zn phase diagram becomes more complex, but the changes are such as to help the technologist rather than hinder him. Zinc arsenides form and at 700 °C, for instance, two regions exist on the diagram which involve condensed phases (Casey and Panish 1968). These regions contain respectively, GaAs, $ZnAs_2$ and Zn_3As_2 (all solids), and GaAs, Zn_3As_2 and a liquid containing all three elements. It has been shown in section 3.1 that such a region is invariant, i.e. all diffusions carried out on systems within the same region should produce the same result. The use of invariant regions in Zn/GaAs diffusions has been discussed at some length by Casey and Panish. They recommend using the first of the two regions, involving three solids, since a liquid can cause a deterioration of the surface, if that is where it happens to form.

4.3 Zinc in GaP

4.3.1 Diffusion mechanism

Radiotracer diffusion profiles for zinc in GaP have been produced by Chang and Pearson (1964a, 1964b) over the temperature range 800 °C–1100 °C and by Luther and Wolfstirn (1973) at 850 °C. The largest body of experimentation of this type, however, was reported by Tuck and Jay (1976, 1977a, 1977b), who worked at 900 °C. Typical results from their work are shown in figure 4.8. The curves show a concave section, resembling profiles for zinc in GaAs, and certainly do not correspond to any of the simple Fick's law solutions. This suggests that the diffusion mechanism is complex and that the diffusion coefficient, D, might be a function of concentration. This possibility was pursued by Chang and Pearson (1964b), who carried out isoconcentration experiments at 1000 °C. Good erfc curves were found for three different concentrations; their results are plotted in figure 4.9. The line through the points corresponds to a diffusion coefficient proportional to the square of concentration. Chang and Pearson plotted a rather more complicated function, but this hardly seems justified on the basis of only three points. The marked similarity between this result and the corresponding Zn/GaAs result (see, for example, figure 4.2) strongly suggests that some substitutional–interstitial mechanism applies in both cases and this conclusion was, in fact, drawn by Chang and Pearson.

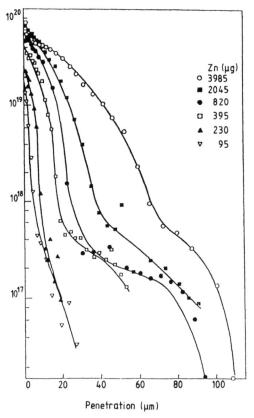

Figure 4.8 Zinc profiles in GaP after diffusion at 900 °C, with different amounts of zinc in diffusion ampoules (from Tuck and Jay 1977a).

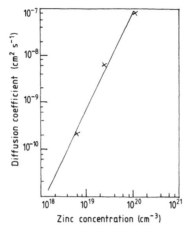

Figure 4.9 Diffusion coefficient of zinc in GaP at 1000 °C
as a function of zinc concentration (data taken from Chang
and Pearson 1964).

4.3.2 Phase diagram considerations

The ternary phase diagrams for Ga–As–Zn and Ga–P–Zn show
the same qualitative features. Above about 900 °C and 1050 °C
respectively, the diagram is of the form of figure 4.5. Below these

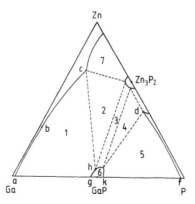

Figure 4.10 Isothermal section of the gallium–phosphorus–zinc phase diagram for 900 °C. The GaP and
Zn₃P₂ solid regions are shown enlarged.

temperatures, zinc solids are created and a more complicated form emerges, shown for Ga–P–Zn at 900 °C in figure 4.10 (Panish 1966b). In practice, diffusions in GaP will normally take place below 1050 °C, so figure 4.10 is the more significant phase diagram. The discussion given in this section may be taken as complementary to that already presented in section 4.2.3.

In figure 4.10 the liquidus lines abc and df are drawn accurately, taken from Panish's paper. The solidus line for GaP, ghk, and also the solidus for the Zn_3P_2 are drawn enormously exaggerated, however, for the sake of making the following commentary more clear. Similarly, the width of region 3 is shown much exaggerated in the diagram. The equilibrium represented by the various regions may be described as follows.

Region 1

A system described by a point in this region is made up of solid GaP, a liquid and a vapour phase. The composition of the liquid is given by a point such as b on the liquidus line, and the composition of the solid is given by a point such as h on the solidus. Corresponding points on the solidus and liquidus curves can be joined by a tie-line bh. In this system there are therefore three components and three phases. The number of degrees of freedom is given as two by the Phase Rule, one of which has already been used in choosing the temperature to be 900 °C. The other degree of freedom can be used to change the conditions in the system without its leaving region 1. If the weights of the elements are changed slightly, the point representing the system moves to another position in the region. The point b moves to another place on the liquidus, b', say, and h moves to h'. The system now has a new tie-line, b'h', and the compositions of all three phases in the system have changed.

Region 2

In this region four phases co-exist (liquid c, vapour, solid Zn_3P_2, solid GaP) and the Phase Rule indicates that there are no degrees of freedom left after setting the temperature, i.e. the system is invariant and causing the point representing the system to move about within the region does not change the compositions of the phases; all that happens is that the relative amounts of the phases change.

Region 3
This is a three-phase region in which two of the phases are solids. If it were drawn accurately its width would be much too small to show on a diagram such as figure 4.10. It will be demonstrated below, however that the region can be of some importance in diffusion work.

Region 4
This is similar to region 2. The composition of the liquid is given by the point d rather than by c.

Region 5
This is similar to region 1 except that the composition of the liquid is given by a point on the liquidus df.

Region 6
A system within region 6 is composed of solid GaP surrounded by its associated vapour. There are thus two degrees of freedom available to the experimenter after setting the temperature. This phase is very small and is usually represented by a point on diagrams such as figure 4.10. (The same applies to the area inside the Zn_3P_2 solidus). From the point of view of diffusion, however, the region is of importance and many diffusion conditions put the system here; in passing it might be added that this comment applies equally to the Zn/GaAs phase diagram of figure 4.5, in which the solid region is shown, correctly to scale, as a point.

Region 7
To make the system enter this region it would be necessary to use a weight of zinc in excess of the weight of GaP. The region is not likely to be of importance in semiconductor diffusion work.

Consider a diffusion experiment in which a slice of GaP is contained inside an ampoule together with a piece of zinc and, perhaps, some excess phosphorus. To fix ideas, let the size of the ampoule be 6 cm^3; this was the size used by Tuck and Jay (1976, 1977b) in their $900\,^{\circ}$C experiments. Consider initially experiments in which no phosphorus is added to the diffusion ampoule. If no zinc is added either, i.e. if we just heat up a piece of GaP in a closed ampoule at $900\,^{\circ}$C, more phosphorus than gallium enters the vapour phase (Jordan 1971) and the condensed phases become gallium-rich. The point representing the system therefore moves to the left along the Ga–P baseline and comes to rest at some point between a and g. We then have in the ampoule a liquid phase of composition a and a solid of composition g. If zinc is then added, the point moves off the baseline and enters region 1. When sufficient zinc has been added, the system-point will be on the tie-line bh. It is important to realise that the point will normally be much nearer to h than to b because, in general, the weight of GaP in the ampoule will be much greater than the weight of zinc.

In order to calculate the exact position in region 1 for a given diffusion, it is necessary to know the form of the liquidus abc and the variation of zinc and phosphorus vapour pressures as the system moves along abc. The first of these requirements is given in figure 4.10, and the second is contained in a paper by Jordan (1971). Strictly, it is also necessary to know the vapour pressure of gallium along abc. As with the GaAs case, however, it is so small that it can safely be ignored. The calculation is performed using the same iterative technique that was described for Zn/GaAs in section 4.2.3.

As more zinc is added to the ampoule, the tie-line moves up the liquidus and eventually reaches the point c. If yet more zinc is added, the system then enters region 2. Further addition of zinc makes no difference to the compositions of the phases in the ampoule; it merely goes towards making more liquid c and Zn_3P_2. The gallium and phosphorus required for these last two processes come from the GaP slice. The zinc vapour pressure in region 2 is 0.65 atm (Jordan 1971). In a $6\,cm^3$ ampoule this requires about 3 mg zinc in the vapour phase. The phosphorus vapour pressure in region 2 is 1.5×10^{-3} atm and is almost all in the form P_2. This amounts to about 8 μg of phosphorus in the vapour phase.

If phosphorus is now added to the ampoule, the system-point moves to the right in figure 4.10, initially staying within region 2. Eventually, it moves into region 3 and, if sufficient phosphorus is added, into region 4. In this latter region the phosphorus vapour pressure is about 3 atm (Nygren and Pearson 1969) and mostly in the form P_4. This amounts to having approximately 23 mg of phosphorus in the vapour phase in a $6\,cm^3$ ampoule. Thus although region 3 is very narrow on the phase diagram it does play rather a crucial role. As the system moves left to right across it, the weight of phosphorus in the vapour changes by several orders of magnitude. The amount of zinc also changes, but by a smaller factor, reducing to 0.18 atm in region 4. Diffusions carried out in region 4, therefore, require just a little more care in order to ensure that the exploding ampoule phenomenon is not introduced into the experiment. This is even more true of region 5, which is probably best avoided.

If the ampoule contains phosphorus with a very small amount of zinc, the system point is likely to be in region 6 of the phase diagram. In order to understand this, it is necessary to return to the simple experiment of heating a slice of GaP in a closed ampoule at $900\,^{\circ}$C. The system inside the ampoule is represented by a point on the gallium side of the base-line, between a and g. If phosphorus is added, the system moves to the right along the base-line and, if the amount is more than a few μg, it moves into the solidus region (between g and k). A very small amount of the added phosphorus goes to adjust the stoichiometry of the GaP crystal, but most of it goes straight into the vapour phase. If sufficient phosphorus is added, the system can move beyond k into region 5. However, the quantity of added phosphorus would need to be larger than would normally be used in this type of experiment.

Consider the enlargement of the solidus shown in figure 4.11 and assume that sufficient phosphorus has been added to take the system to point l. Now add a small amount of zinc without changing the weight of phosphorus. A very small proportion of the zinc goes into the solid and all of the rest goes into the vapour phase, since there is no liquid formed in region 6. Thus, to a good approximation, all of the added zinc and phosphorus enters the vapour. It is therefore a simple matter to calculate the vapour pressures for these experiments. The fraction of phosphorus in the solid phase is fixed by the external phosphorus pressure. Since the phosphorus pressure does not change when zinc is added, the system must therefore move away from the point l along a line of constant phosphorus content l m. If the weight of zinc is increased, the system eventually arrives at the solidus boundary, i.e. the point m. In figure 4.11 this point is shown at the edge of region 3. The further addition of zinc causes the system to move into region 3, and some of the zinc must then be used to form Zn_3P_2. This has the effect of removing phosphorus from the vapour, causing the system to move rather sharply into region 2, where it remains as yet more zinc is added. By this time, of course, at least 3 mg of zinc would have been put into the diffusion ampoule.

In principle, the system can also be taken along a constant phosphorus line such as no, ending in region 1, or along a line such as pq, ending in region 5. In practice, neither of these would be easy to achieve. The amount of phosphorus required to put the system on the line no would have to be very small, of the order of a few μg. Conversely, more than 23 mg of phosphorus would be needed to start at a point such as p. While this is quite possible, it might be considered unwise, for reasons noted above.

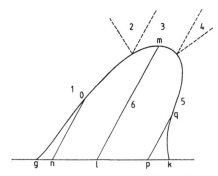

Figure 4.11   The GaP/Zn solidus (from Tuck and Jay 1976).

4.3.3   *Application to diffusion conditions*

Tuck and Jay (1977b) diffused zinc into n-type GaP at 900 °C over a wide range of diffusion conditions. Before diffusion, the slices were carefully polished to produce a mirror finish on both surfaces. At the end of the diffusion process, the surfaces were inspected under the microscope and pits and other imperfections were observed. The samples were then sectioned and chemical staining techniques were used to reveal the p–n junctions. The depth of each junction was measured and the junction planarity was assessed.

Using the techniques outlined in the previous section, the system-point was determined on the ternary phase diagram for each diffusion. In all, 49 diffusions were reported in regions 1,2,3 and 6. They found that samples with badly pitted surfaces tended to have poor, non-planar junctions, the correlation proving to be strongest in samples with relatively shallow junctions. It was suggested that the likely sequence of events in producing a 'poor' junction is as follows. The initially flat sample loses material in helping to set up the thermodynamic equilibrium which is formed in the ampoule. This causes pitting of the surface. The diffusion then takes place from an irregular surface, leading to irregularity in both the diffusion front and the p–n junction. The closer the junction is to the surface, the more closely does it follow the surface irregularities; in deeper diffusions, the irregularities tend to iron out.

The predictions made in the previous section concerning invariant regions were borne out by this work; the diffusions which took place in region 2 of the phase diagram all gave the same junction depth even though the amounts of zinc and phosphorus in the ampoules varied over quite a large range. It was clear, however, that one important parameter does vary within an 'invariant' region. The amount of material removed from the specimen depends on the position of the system-point inside the region and this, in turn, affects the quality of the surface and of the junction. For this reason, it was found to be difficult to produce high-quality surfaces and planar junctions when working in region 2. The diffusions which took place in region 6, on the other hand, produced consistently good surfaces and junctions; this is not surprising in view of the fact that in this region very little material is removed from the GaP in setting up the equilibrium. In addition, there is no liquid within the ampoule and this also helps to maintain surface quality.

Similar experiments were carried out by Nygren and Pearson (1969), who confined their experiments to regions 2 and 4. They also found difficulty in obtaining good results from region 2 diffusions, although they were of the opinion that the non-planarity of their junctions was due primarily to the presence of diffusion-induced dislocations. Their region 4 diffusions normally gave better results. This would be expected on the basis of the above argument, since the amount of material taken from the GaP slice in setting up the ampoule equilibrium is much less when region 4 is used.

The lesson from all this would seem to be that when carrying out diffusions of zinc in GaP, regions 4 and 6 are to be preferred. Region 6 gives good results most of the time, but it is not easy to obtain reproducibility, since the region is not invariant. Good reproducibility can be obtained using region 4, but the experimenter has to be prepared to accept diffusion ampoules containing

3 atm of phosphorus vapour. Another alternative is to use region 2 but to take steps to preserve the quality of the GaP surface. Various tricks of the trade have been developed. They include having powdered GaP in the ampoule in the hope that this, rather than the slice, will be used to set up the equilibrium conditions, using layers of oxide or nitride to protect the surface, and having two polished layers face to face in the ampoule. These methods work sometimes.

4.4   Zinc in InP

Zinc appears to diffuse differently in InP. Work by Kundukhov *et al* (1967) suggested a large discrepancy can exist between the concentrations of zinc atoms and holes in a zinc-diffused specimen. This result was checked by Tuck and Zahari (1977a) who, by using very long diffusion times, were able to prepare InP homogeneously doped with radioactive zinc. Standard electrical measurement techniques were then used to make a direct comparison between atom and hole concentrations. It was found that at the highest concentrations the atomic concentration exceeded that of holes by almost two orders of magnitude. Most of the zinc therefore exists in some form other than a simple acceptor on the indium site.

Four radiotracer profiles were presented in a paper by Chang and Casey (1964) for temperatures in the range $600\,°C$–$900\,°C$. The curves looked rather like profiles of zinc in GaP, showing the familiar concave section, and Chang and Casey proposed a similar substitutional–interstitial mechanism. They analysed the profiles using the Boltzmann–Matano method and came to the conclusion that diffusion coefficient was a function of zinc concentration.

A rather more extensive radiotracer investigation was reported by Tuck and Hooper (1975), also diffusing from an external vapour. This work revealed substantial differences between diffusion in InP and the other III–Vs. The surface concentrations of the profiles could be very high, for instance, exceeding $10^{21}\,\text{cm}^{-3}$ in some cases. Most significantly, however, it was found that an increase in the ambient phosphorus pressure brings about a decrease in the surface concentration of zinc, indicating that most of the zinc does not occupy the 'conventional' group III site. This can be shown as follows. Consider the equilibrium at the surface between zinc atoms in the vapour phase and those on indium sites in the crystal:

$$Zn_{vap} + V_{In} \rightleftharpoons Zn_{In} \qquad (4.40)$$

which, using the law of mass action, leads to

$$P_{Zn}[V_{In}] = K_1[Zn_{In}] \qquad (4.41)$$

where P_{Zn} is the zinc vapour pressure, $[V_{In}]$ represents the concentration of indium vacancies and $[Zn_{In}]$ is the concentration of zinc atoms on indium sites. K_1 is a constant. At fixed temperature

$$[V_{In}] = K_2 P_{P_2}^{1/2} \qquad (4.42)$$

(cf. equation (3.3)) where it is assumed that phosphorus exists in the vapour phase in the diatomic form. Hence from equations (4.41) and (4.42)

$$K_2 P_{Zn} P_P = K_1[Zn_{In}]. \qquad (4.43)$$

Thus if most of the zinc occupied indium sites, an increase in phosphorus pressure would bring about an increase in surface concentration. Most of the zinc therefore appears in the lattice in a less simple form than might have been expected. Tuck and Hooper suggested the neutral complex $(V_P Zn_{In} V_P)$. This is similar to the model proposed by Young and Pearson (1970) for sulphur in GaAs and described in section 3.1. It is not identical, however, since the impurity atom occupies the group III site in this case, rather than the group V. The equilibrium between the zinc in the external vapour and the zinc in the crystal becomes

$$Zn_{vap} + V_{In} + 2V_P \rightleftharpoons V_P Zn_{In} V_P \qquad (4.44)$$

giving

$$[V_P Zn_{In} V_P] = K_3 P_{Zn}[V_P]^2[V_{In}]. \qquad (4.45)$$

At equilibrium, we have

$$[V_{In}][V_P] = K_4 \qquad (4.46)$$

(cf. equation (3.26)). In addition, it follows from equations (4.42) and (4.46) that

$$[V_p] = \frac{K_5}{P_{P_2}^{1/2}}. \qquad (4.47)$$

Using equations (4.46) and (4.47) in equation (4.45),

$$[V_P Zn_{In} V_P] = K_3 K_4 K_5 \frac{P_{Zn}}{P_{P_2}^{1/2}} \qquad (4.48)$$

which, at least qualitatively, gives the dependence on phosphorus pressure which was found experimentally. Some fraction of the zinc must appear on the conventional site, of course, otherwise the zinc-doped material would not be p-type. We have, for charge neutrality,

$$Zn_{In}^- \simeq p. \qquad (4.49)$$

The concentration of holes p is therefore much less than the total concentration of zinc, in agreement with the electrical measurements noted above.

Tuck and Hooper suggested that the complex is relatively immobile (note that this is another difference between their model and that of Young and Pearson) and that the remaining zinc diffuses by the conventional substitutional–interstitial mechanism. This means that a small proportion of the zinc atoms on indium sites leave to go interstitial and travel through the lattice with very high diffusivity. At some point inside the crystal an interstitial finds an indium vacancy and joins the lattice again. The diffusion mechanism is illustrated in figure 4.12. Suppose further that the diffusing interstitial is uncharged. The equilibrium between the complex and the substitutional zinc at any point in the crystal can be written

$$V_P Zn_{In} V_P \rightleftharpoons 2V_P + Zn_{In}^- + p. \qquad (4.50)$$

Similarly, that between the substitutional and interstitial zinc is

$$Zn_{In}^- + p \rightleftharpoons Zn_i + V_{In}. \qquad (4.51)$$

If it is assumed that instantaneous local equilibrium exists so that the law of mass action can be used, equations (4.50) and (4.51) give

$$[V_P Zn_{In} V_P] = K_6 [V_P]^2 [Zn_{In}^-]^2 \qquad (4.52)$$

and

$$[Zn_{In}^-]^2 = K_7 [Zn_i][V_{In}] \qquad (4.53)$$

where equation (4.49) has been used to eliminate p. Removing

$[Zn_{\overline{In}}]$ between equations (4.52) and (4.53) gives

$$[Zn_i] = \frac{[V_P Zn_{In} V_P]}{K_8 [V_P]} \qquad (4.54)$$

where $K_8 = K_4 K_6 K_7$.

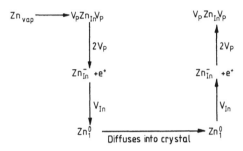

Figure 4.12 Proposed diffusion mechanism for zinc in InP (from Tuck and Hooper 1975).

Call the zinc concentration at any point in the crystal [Zn]; this will be a function of x, of course. We now use the assumptions that the total zinc concentration is approximately equal to that of the complex, i.e. $[Zn] \simeq [V_P Zn_{In} V_P]$, and that the diffusion coefficient of interstitials is very large compared to those of the substitutional zinc and the complex. The diffusion coefficient, by analogy with equation (3.13), is given by

$$D = D_i \frac{\partial [Zn_i]}{\partial [Zn]} \qquad (4.55)$$

where D_i is the interstitial diffusion coefficient, i.e.

$$D = \frac{D_i}{K_8 [V_P]}. \qquad (4.56)$$

The model therefore predicts a diffusion coefficient which at thermodynamic equilibrium is constant, independent of zinc concentration. A normal chemical diffusion would not be expected to occur at thermodynamic equilibrium, however; we would expect the vacancy equilibrium to break down in just the same way that it does when zinc diffuses in GaAs and GaP. This argument would seem to account for the similarity in the shapes of the profiles from all three semiconductors, each showing the characteristic concave section.

It is possible to check equation (4.56) using isoconcentration techniques and this was subsequently done by Tuck and Zahari (1977b), working at 800 °C. Their result is shown in figure 4.13. The diffusion coefficient proves to be constant over almost two orders of magnitude; the contrast between this result and the equivalent one for GaAs shown in figures 4.2 and 4.3 is really quite striking.

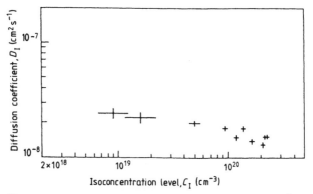

Figure 4.13   Diffusion coefficient of zinc in InP at 800 °C as a function of zinc concentration (from Tuck and Zahari 1977b).

A rather more complex proposal was put forward by Yamada *et al* (1983) for zinc in InP. They explained the anomalous 'double-hump' shape of the profile by dividing the InP sample into two regions and defining quite different diffusion regimes for them. In this model, the region close to the surface has most of the zinc on indium sites; diffusion is by interstitials, as in the normal substitutional–interstitial theory. Further into the crystal, most of the zinc is tied up in the complex $(Zn_{In}V_P)$ which is immobile. Diffusion occurs by the complex $(V_PZn_{In}V_P)$, which is here assumed to be mobile. These systems are assumed not to interact in any way, so each gives rise to an independent profile. The predicted profile is simply the sum of the two, hence its 'double' shape. In order to match their SIMS profile, taken after a 700 °C diffusion, it was necessary to assume that the $(V_PZn_{In}V_P)$ complex diffuses almost three times as fast as the zinc interstitial. This must be seen as a weakness in the theory, as must the requirement of two parallel diffusion processes proceeding without any interaction. Further experimentation would be necessary to fully evaluate their theory.

As a post-script to this description of models for the Zn/InP system, it is worth noting work by Dlubek *et al* (1985), who carried out positron lifetime measurements on zinc-doped InP crystals. This technique is able to detect vacancy defects in metals, alloys and semiconductors. Dlubek *et al* found evidence for divacancies in approximately the same concentration as uncharged zinc atoms. As they point out, this ties in very well with the postulate of an uncharged ($V_P Zn_{In} V_P$) complex, which is common to both of the above theories.

4.5 Zinc in Other Compounds

Showan and Shaw (1969) reported both chemical and isoconcentration experiments in AlSb, using radioactive zinc. The chemical diffusion profiles were taken in the temperature range 650 °C–933 °C. The form of the curves resembled those taken by other workers for zinc in GaAs. Boltzmann–Matano analysis was performed on the profiles but, as with the Zn/GaAs case, the results were indeterminate, different curves giving different values of diffusion coefficient D for the same zinc concentration. It seems likely that the reason is the same as that given above for Zn/GaAs; the defect equilibrium broke down during diffusion. Only a small number of isoconcentration measurements were reported, all at 933 °C. These gave the expected erfc profiles and values of D in the range 8×10^{-9}–6×10^{-8} cm^{2}s^{-1} for zinc concentrations of 6.5–9.7×10^{19} cm^{-3}. The number of profiles taken was too small to permit any relationship to be determined for diffusion coefficient as a function of concentration. It is clear from this data, however, that D depends quite strongly on concentration of zinc. This finding, together with the shape of the chemical diffusion profiles, suggests that the substitutional–interstitial mechanism operates in the system.

The isoconcentration approach was also used by Boltaks *et al* (1969), studying diffusion of zinc in InAs. They produced six radiotracer profiles, all taken at 800 °C, for isoconcentration levels varying from 5×10^{17} cm^{-3} to 5×10^{19} cm^{-3}. Their results are plotted in figure 4.14. It can be seen that above about 5×10^{18} cm^{-3}, the diffusion coefficient increases approximately as the square of concentration, but that for lower values, the coefficient is constant. Both of these characteristics are consistent with a substitutional–interstitial interpretation for the diffusion mechanism. InAs is a low band-gap semiconductor and in consequence has a relatively high value of intrinsic carrier concentration, n_i. The concentration 5×10^{18} cm^{-3} corresponds quite well to that of intrinsic carriers at 800 °C given by Casey (1973). For concentrations below n_i, the neutrality condition which was used in obtaining equation (4.3) is replaced by $p = n_i$ and equation (4.2) then becomes

$$K_1 n_i^2 S = IV. \tag{4.57}$$

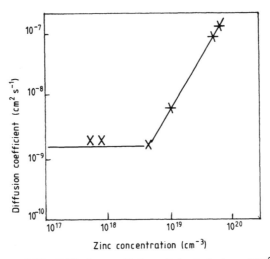

Figure 4.14   Diffusion coefficient of zinc in InAs at 800 °C as a function of zinc concentration (data taken from Boltaks *et al* 1969). Reproduced by permission of *Am. Inst. Phys.*

The argument of section 4.1 then leads to the net diffusion coefficient being constant, as found by Boltaks *et al*. For zinc concentrations in excess of n_i the sample becomes p-type and the normal neutrality condition applies. The diffusion coefficient then increases with concentration, as in the Zn/GaAs case. Investigations giving broadly similar results have been carried out for InSb (Kendall 1968, Mozzi and Lavine 1970).

Da Cunha and Bougnot (1974) presented chemical diffusion profiles of radioactive zinc in GaSb in the temperature range $560\,°C–640\,°C$. They also measured the profile of holes for a sample diffused at $610\,°C$. In many respects, their results resemble those reported above for InP. The profile shapes were similar and some of them had very high values of surface concentration, in the region of $10^{21}\,cm^{-3}$. The hole concentration, also, was found to be substantially less than the concentration of zinc atoms. Two isoconcentration profiles were taken at $560\,°C$ and $580\,°C$. Values for D of $1.8 \times 10^{-11}\,cm^2\,s^{-1}$ and $8 \times 10^{-11}\,cm^2\,s^{-1}$ were found, respectively, for isoconcentration levels of $4.5 \times 10^{20}\,cm^{-3}$ and $6 \times 10^{20}\,cm^{-3}$. It is fairly clear from this work that D is a function of concentration, but more experimentation is required before any more firm conclusions can be drawn.

There are very few reliable studies in the III–V solid solutions. Yamamoto and Kanbe (1980) used p–n junction measurements to study the diffusion of zinc into $Ga_{1-x}In_xAs$ for the whole range of x and temperatures $525\,°C–600\,°C$. They analysed their results using a number of models and suggested that the substitutional-interstitial mechanism was probably operating with a doubly charged zinc interstitial. As noted in a previous chapter, the p–n junction method is a rather crude technique for studying diffusion, so some care should be taken in interpreting data of this sort. Blum *et al* (1983) produced very shallow diffusions in $Ga_{1-x}Al_xAs$ using radiotracer zinc. Their experiments were carried out at $600\,°C$ and the samples were produced by liquid phase epitaxy. A source of composition Zn_9GaAs_{10} was used. For the $x = 0$ (i.e pure GaAs) experiments, the samples had a range of dislocation densities and were initially doped either with germanium, silicon or tin, giving respectively p-, n- and n-type material. In all cases the doping level was $1 \times 10^{18}\,cm^{-3}$. It was found that the results were independent both of the starting dislocation density and of background doping; this second finding is perhaps not too surprising in view of the fact that the profiles gave very high values of zinc concentration, rising at the surface to about $10^{21}\,cm^{-3}$. The Boltzmann–Matano method was used to analyse the profile for GaAs and a value of $1.4 \times 10^{-11}\,cm^2\,s^{-1}$ is quoted for a concentration of

1×10^{19} cm^{-3}. Because of the problems associated with this technique, outlined in sections 2.5 and 4.2.1, this should be taken as an order of magnitude value only. For $x > 0$, they obtained the interesting result that a change in diffusion profile shape appeared at about $x = 0.1$. Below this figure, the profiles resembled those found in GaAs. Above it, a strong surface peak appeared; further increases in x produced little change in the shape of the profiles.

4.6   Cadmium in InP

Considerable interest has been shown in the use of cadmium as a p-type dopant in InP. Tien and Miller (1979) carried out a series of diffusions at four temperatures in the range 581 °C–731 °C. P–n junction depths were measured in n-type samples with background doping covering 2×10^{16}–9×10^{18} cm^{-3}. In this way, a profile was produced for each of the four temperatures, each profile containing about five points. It was noted by Tien and Miller that these were profiles of ionised cadmium concentration and not atomic profiles, a distinction which is important in the light of the theory they put forward to explain the results. They used the same diffusion source for all experiments so that all diffusions were carried out with an excess overpressure of phosphorus. In a second series of experiments, at 681 °C, diffusions were performed with different amounts of phosphorus in the diffusion ampoule. They found that increasing the weight of phosphorus had the effect of decreasing the junction depth. On the basis of these results and an assumed similarity between the Zn/InP and Cd/InP systems, the following theory was proposed.

The assumption was made that the cadmium can exist in the lattice in three forms: an ionised acceptor Cd_{In}^{-}, a neutral complex involving a cadmium atom on an indium site and as an unionised interstitial, Cd_i. It is further assumed that for high cadmium concentrations most of the dopant is in the complex form. The interstitials, although only present in low concentration, can diffuse rapidly; both the other forms are immobile. This theory is very similar to that proposed by Tuck and Hooper (1975) for zinc in InP, the main difference being that Tien and Miller do not specify the exact form of their complex. The equilibrium condition between the interstitials and the acceptors is

$$Cd_{In}^- + h \rightleftharpoons Cd_i + V_{In} \qquad (4.58)$$

where h is a hole. If we simplify the nomenclature by calling the concentrations of interstitials, substitutionals and indium vacancies C_i, C_S, C_V respectively, the mass action law for equation (4.58) can be written

$$pC_S = KC_iC_V. \qquad (4.59)$$

For charge neutrality, we have $p = C_S$, so

$$C_S^2 = KC_iC_V. \qquad (4.60)$$

Since the concentration of interstitials is negligibly small, the continuity equation for atom transport is

$$\frac{\partial C_n}{\partial t} + \frac{\partial C_s}{\partial t} = \frac{\partial}{\partial x}\left(D_i \frac{\partial C_i}{\partial x}\right) \qquad (4.61)$$

where C_n is the concentration of the neutral complex and D_i is the diffusion coefficient for the interstitials. Using

$$\frac{\partial C_n}{\partial t} = \frac{\partial C_n}{\partial C_s}\frac{\partial C_s}{\partial t} \qquad (4.62a)$$

and

$$\frac{\partial C_i}{\partial x} = \frac{\partial C_i}{\partial C_s}\frac{\partial C_s}{\partial x} \qquad (4.62b)$$

we obtain

$$\left(\frac{\partial C_n}{\partial C_s} + 1\right)\frac{\partial C_s}{\partial t} = \frac{\partial}{\partial x}\left(\frac{2D_iC_s}{KC_V}\frac{\partial C_V}{\partial x}\right) \qquad (4.63)$$

This equation was solved numerically for C_S and they found that they could fit the experimental points for all four temperatures to the same 'normalised' profile. In order to do this, however, it was necessary to choose an appropriate value for the quantity $D^0 = 2D_i/KC_V$ for each temperature and to assume an empirical relationship between neutral and substitutional concentrations of $C_n = 40C_S^{3.5}$ for all temperatures. This normalised curve is shown in figure 4.15; the values of D^0 are given in figure 4.16. Inspection of the right-hand side of equation (4.6.3) suggests that an increase in C_V (i.e. in indium vacancy concentration) should decrease the effective diffusion coefficient. This is in agreement with the experimental finding, noted above, that increasing the phosphorus pressure in the diffusion ampoule decreased the measured junction depth.

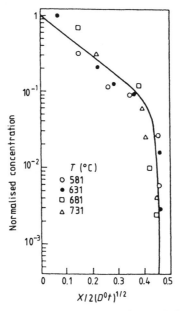

Figure 4.15 'Normalised' profile for cadmium in InP (from Tien and Miller 1979).

Similar studies were carried out by Ando *et al* (1981) on the diffusion of both cadmium and zinc into InP. They used both the p–n junction method and the capacitance–voltage technique to produce profiles of ionised acceptors. The results were broadly similar to those of Tien and Miller, and the authors believed that they could be explained by the same model.

A rather different model was put forward by Kuebart *et al* (1983) to account for their $C-V$ measurements on cadmium-diffused material. They proposed that cadmium can exist in the lattice either as a singly or doubly ionised acceptor and that these two forms have different diffusion coefficients. The model, which is rather similar to an early one for the diffusion of zinc in GaAs (Allen 1960), allows for the built-in field effect which occurs in the diffused region due to the acceptor gradient. A number of fitting parameters were employed to obtain correspondence between the theory and the experimental profiles. No allowance is made in the theory for the possibility of some of the cadmium atoms being uncharged and any such effect would not have been revealed by their experiments. Much more experimentation is required before the model can be fully assessed; it would certainly not fit the experimental results reported for the related Zn/InP system.

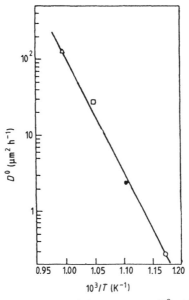

Figure 4.16   Variation of the parameter D^0 with temperature (from Tien and Miller 1979).

Metallurgical studies on cadmium-diffused InP were carried out by Dutt *et al* (1981) and by Dutt and Brasen (1983). They performed diffusions at 680 °C for a range of ambient cadmium and phosphorus vapour pressures. Depths of p–n junctions were measured and plotted as functions of vapour pressure. Because of the rather limited nature of this type of evidence, they were forced into making some rather gross assumptions in interpreting the results. These included putting the measured junction depth, x_j, proportional to $(Dt)^{1/2}$, where D is diffusion coefficient and t is time of diffusion, and assuming that if more than one diffusion process was operative, they did not interact. The diffused samples were also studied using transmission electron microscopy. The authors suggested that the diffusion mechanism depends on wither the experiment takes place on the indium-rich or the phosphorus-rich side of the phase diagram. On the indium-rich side, they proposed that a substitutional–interstitial mechanism operates with a neutral interstitial species. On the phosphorus-rich side, the mobile interstitial atoms are doubly ionised and much of the cadmium is incorporated in the crystal in phosphorus precipitates. Supporting evidence for this last proposal is given by their electron microscopy, which showed precipitation in samples diffused under high ambient vapour pressure of phosphorus.

On the whole, our knowledge of this important system is rather unsatisfactory. Interesting models have been put forward for the diffusion mechanism but these have been supported by inadequate experimentation. Virtually all of the experiments which have been reported have involved measuring p–n junction depths and, as has been noted several times, this method is too crude to study complex diffusion mechanisms. The general assumption, sometimes stated, sometimes not, that the Cd/InP system behaves just like the Zn/InP must be seen as risky.

4.7 Cadmium in Other Compounds

The diffusion of cadmium into AlSb has been studied by Shaw and Showan (1969b), using radiotracer cadmium. Four profiles were presented, all taken at $900\,^{\circ}C$ for different vapour pressures of cadmium in the diffusion vessel (see figure 4.17). The profiles showed the sharp diffusion front which is characteristic of systems for which diffusion coefficient is a function of concentration. In addition they all showed the 'tails' which are often reported for group II diffusants in III–V semiconductors (only one is shown in figure 4.17). They are usually attributed to enhanced diffusion along dislocations. Two isoconcentration experiments were performed at the same temperature for different isoconcentration levels. Diffusion coefficients of $8.0 \times 10^{-10}\,cm^2\,s^{-1}$ and $2.85 \times 10^{-10}\,cm^2\,s^{-1}$ were found for isoconcentration levels of $1.6 \times 10^{19}\,cm^{-3}$ and $1.3 \times 10^{19}\,cm^{-3}$ respectively. Hall measurements made on the two homogeneously doped samples revealed reasonable agreement between cadmium and hole concentrations, indicating that cadmium acts primarily as a singly ionised acceptor in the AlSb lattice. Shaw and Showan came to the conclusion that the diffusion mechanism was a substitutional–interstitial process with a singly charged interstitial atom. They also saw evidence that a breakdown of defect equilibrium occurs in this system in much the same way as in the Zn/GaAs system.

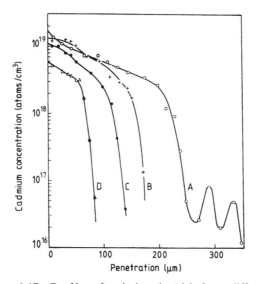

Figure 4.17 Profiles of cadmium in Alsb for a diffusion temperature of 900 °C, all profiles normalised to a diffusion time of 100 h (from Shaw and Showan 1969b). Reproduced by permission of *Akademic-Verlag, Berlin.*

Arseni *et al* (1967) produced four radiotracer profiles of cadmium in InAs, for temperatures of 650 °C, 750 °C, 800 °C and 900 °C. They were able to fit erfc curves to all of them, giving values of D varying from 4×10^{-10} cm^{2}s^{-1} to 8×10^{-9} cm^{2}s^{-1}, following the relationship

$$D = 7.4 \times 10^{-4} \exp\left(-\frac{1.15 \ (eV)}{kT}\right) \ cm^2 s^{-1}. \qquad (4.64)$$

This result led them to propose that cadmium diffuses by a vacancy mechanism in InAs. P–n junction depths were measured by Zotova *et al* (1967), who used n-type InAs substrates of doping 4.7×10^{16} cm^{-3}–3.7×10^{19} cm^{-3}. It was shown that adding excess arsenic to the diffusion ampoule decreased the penetration of cadmium; this is the result which would be expected if the dopant were diffusing by a substitutional–interstitial mechanism. None of these workers measured both the concentration of holes and that of cadmium atoms, so it is not clear whether cadmium acts as a simple acceptor in this material.

Radiotracer work was also carried out by Orth and Watt (1965) in studying the Cd/InSb system. Their method was to electroplate a thin layer of radioactive cadmium to the top surface of the semiconductor slice, thereby approximating the 'thin film' boundary condition. The profiles followed a Gaussian curve reasonably well, but the deeper experimental points did not lie on the line. Within the temperature range 434 °C–519 °C, the diffusion coefficient was found to follow the form

$$D = 1 \times 10^{-9} \exp\left(\frac{0.52 \text{ (eV)}}{kT}\right) \text{ cm}^2 \text{ s}^{-1}. \tag{4.65}$$

Some work on cadmium diffusion in InSb by Kolodny and Shappir (1978) is worth a mention for the novelty of the technique used. Cadmium was diffused into n-type material at 400 °C and the profile of ionised acceptors was measured by first polishing the diffused sample at a small angle to the surface and then making MOS structures on the polished surface. Capacitance–voltage measurements were used to obtain the required profile. The quantity of results is too small, however, to come to any firm conclusions about diffusion mechanism. The assumption by Kolodny and Shappir that they were measuring an atomic profile is also somewhat suspect.

4.8   Diffusion of Other Group II Elements

Diffusion of beryllium into GaP was studied by Ilegems and O'Mara (1972) at 800 °C, 900 °C and 1000 °C. A liquid Ga–Be–P source was employed and profiles were plotted using an atomic absorption technique. The method is similar to the radiotracer technique; layers of known thickness are removed and the quantity of beryllium in each layer is assessed. The sensitivity is inferior to the radiotracer method but is extremely convenient, since no suitable radioactive isotope of beryllium is available. Results are shown in figure 4.18. The profiles are quite similar to those of zinc in the same semiconductor, demonstrating a high surface concentration and deep penetration. It seems likely that the diffusion coefficient is concentration dependent; a Boltzmann–Matano analysis of the 1000 °C curve indicated a coefficient in the region of $10^{-7}\,\mathrm{cm}^2\,\mathrm{s}^{-1}$ for the highest concentrations. In later work, beryllium diffusion in GaAs was studied by preparing buried beryllium-rich layers, grown by molecular beam epitaxy, and subjecting the samples to annealing at 900 °C (Ilegems 1977, McLevige *et al* 1978). The out-diffusion from the highly doped regions was monitored using both SIMS and electrical measurements. Surprisingly, these experiments indicated a very much lower rate of diffusion for GaAs than had been found previously for GaP. Values in the region of $10^{-13}\,\mathrm{cm}^2\,\mathrm{s}^{-1}$–$10^{-14}\,\mathrm{cm}^2\,\mathrm{s}^{-1}$ were reported at 900 °C for beryllium concentrations of about $10^{19}\,\mathrm{cm}^{-3}$.

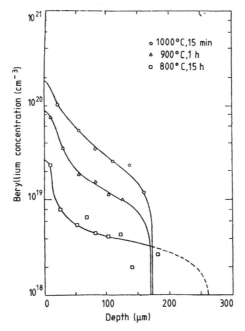

Figure 4.18 Diffusion profiles of beryllium in GaP for three temperatures (from Ilegems and O'Mara 1972).

There is very little data on mercury and magnesium in the III–Vs. Small *et al* (1982) diffused magnesium into GaAs at 835 °C from a liquid source. Profiles were determined using SIMS. They showed a very sharp decrease at the diffusion front, again suggesting a concentration-dependent diffusion coefficient. Penetrations of about 3 μm were obtained for a diffusion time of two hours. Sharma *et al* (1971) used radioactive mercury and diffused into n-type InAs from the vapour phase. At temperatures of 650 °C, 750 °C and 850 °C they obtained profiles which could be fitted by erfc curves. The values of diffusion coefficient varied between $9 \times 10^{-13} \, \text{cm}^2 \text{s}^{-1}$ and $2 \times 10^{-11} \, \text{cm}^2 \text{s}^{-1}$.

Chapter 5

Diffusion of Transition Elements

In the III–V semiconductors, transition elements give rise to deep levels near the centres of the energy gaps (White 1980). The presence of one or more of these elements can have a profound effect on the electrical properties of the material. In GaAs, for example, chromium acts as a deep acceptor and, if used to over-compensate a sample containing shallow donors, can give rise to a crystal with resistivity as high as $10^9 \, \Omega \, \text{cm}$. Iron is commonly used to create the same effect in InP. Material of this type is known as 'semi-insulating'. For a long time it was believed that diffusion of transition elements occurred only very slowly so that, for instance, a conducting layer of GaAs could be grown on a chromium-doped substrate without any risk of the chromium entering the layer. This is now known to be incorrect and the knowledge has had a profound effect on III–V semiconductor technology. A good deal of work has been carried out in recent years with the aim of producing high-resistivity material without resorting to transition elements. In the case of GaAs, this has been largely successful (Farges *et al* 1982, Holmes *et al* 1982, Ta *et al* 1983), although the 'undoped' semi-insulating material is rather more trouble to produce than the chromium-doped. To date there is no report in the literature of a reproducible technique for producing undoped semi-insulating InP. Most of the available diffusion data concerns GaAs and InP and this fact is reflected in the present chapter.

5.1   Chromium in GaAs

In as-grown GaAs, chromium is believed principally to occupy the gallium site, acting as a double acceptor (Brozel *et al* 1978). An early sign that this acceptor might be mobile within the crystal was given by electrical measurements carried out by White *et al* (1976). They reported finding a photocapacitance peak at 0.9 eV, believed to be due to the Cr_{Ga} centre, in epitaxial layers grown on chromium-doped substrates. The possibility therefore arose that the chromium may have diffused out of the substrate into the growing layer. This disagreeable possibility was largely ignored, however, until radiotracer and SIMS work showed quite unambiguously that this was indeed the correct interpretation.

5.1.1   Diffusion from an external phase

Work in which radioactive chromium was diffused into GaAs from the vapour phase over the range of diffusion times 15 min–10 h and temperatures 800–1100 °C was described by Tuck and Adegboyega (1979). The diffusions were carried out inside an evacuated ampoule, about 5 cm^3 in volume, and pure chromium metal was used as the source. The weight of chromium used made no difference to the final result within the range 100–500 μg; this was perhaps not surprising, since the weight of metal required to produce the chromium vapour pressure in the ampoule was in all cases insignificant compared to the weight of the piece used. Interesting changes did occur in the chromium source during diffusion, however. The appearance changed from matt to shiny during the experiment, and the piece increased in weight by up to 100 percent. X-ray microprobe analysis showed that the solid source contained both gallium and arsenic as well as chromium at the end of the experiment. No difference was observed between profiles produced in samples which had originally been n-type and those which had been semi-insulating at the beginning of the anneal.

Typical results are shown in figure 5.1, for a temperature of 1000 °C. The experimental conditions correspond closely to the 'constant surface concentration' case, but it can be seen that the profiles are far from the erfc curves which would be expected from a simple Fick's law interpretation. Each profile shows a high surface peak occupying about 20 μm adjacent to the surface,

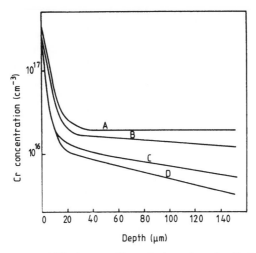

Figure 5.1 Diffusion profiles of chromium in GaAs at 1000 °C. Diffusion times: A, 4 h; B, 3 h; C, 2 h; D, 1 h (from Tuck and Adegboyega 1979).

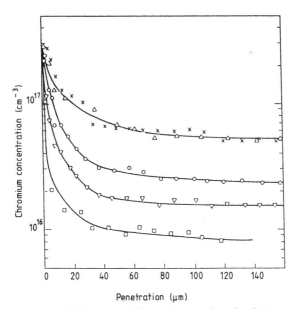

Figure 5.2 Diffusion profiles of chromium in GaAs at 1100 °C: △, 10 h; x, 4 h; ○, 3 h; ▽, 2 h; □, 1 h (from Tuck and Adegboyega 1979).

followed by a region of gentle (or zero) slope which penetrates deep into the crystal. The progression of the profiles with time as revealed by figure 5.1 is interesting. The surface concentration, C_0, is established very early on and remains fixed throughout the experiment; values of C_0 did not change, in fact, between a few minutes and several hours of diffusion time. The bulk concentration, C_B, on the other hand, builds up with time. For lower temperatures and shorter times, the bulk profile is not horizontal. Eventually it becomes so; figure 5.2 shows that at $1100\,^\circ\mathrm{C}$ this occurs after only one hour of diffusion. Once the profile has become horizontal it increases its value slowly as diffusion time is increased. The larger range of times used at $1100\,^\circ\mathrm{C}$ shows, however, that when a concentration of $5 \times 10^{16}\ \mathrm{cm}^{-3}$ is reached, C_B virtually stops rising, so that the value of C_B for a four-hour diffusion is the same as that for a 10-hour one.

Tuck and Adegboyega also investigated the effect of adding small amounts of metallic arsenic to the diffusion ampoule. A set of results taken at $1000\,^\circ\mathrm{C}$ for a diffusion time of four hours and varying quantities of arsenic is shown in figure 5.3. It can be seen that both C_0 and C_B are progressively reduced as the weight of arsenic is increased. In addition, the surface peak penetrates much further into the crystal. This result was rather puzzling. As noted above, it is believed that chromium occupies the gallium site in the GaAs lattice. Increasing the ambient arsenic pressure increases the concentration of gallium vacancies at the surface (see equation 3.3) and this would be expected, in turn, to increase C_0 and then C_B. This argument certainly holds good for zinc diffusion in GaAs, for instance. Zinc also occupies the gallium site and any increase in arsenic overpressure increases the surface concentration of the resulting profile in accordance with the relevant mass-action relation (Tuck 1976). However, as was pointed out by Tuck and Adegboyega, the effect of arsenic pressure on chromium concentration can be completely understood only if the various phases in the ampoule during the anneal are known. During a diffusion, the ampoule contains chromium-doped GaAs, vapours of all three elements and also any condensed phases specified by the phase diagram. It is known that a wide range of Cr–As and Cr–Ga phases can occur, so it seems reasonable to assume that the chromium reacts with the gallium and arsenic to form one or more of these. This assumption is verified by the finding noted above that the source increased in weight and gained both gallium and arsenic.

At any given diffusion temperature the surface concentration of chromium is determined by the vapour pressures of chromium and arsenic in the ampoule. When extra arsenic is added, the composition of the 'chromium phase' and also the vapour pressure will change in a manner determined by the phase diagram. Thus, while the arsenic pressure will increase, it is quite possible that the chromium vapour pressure will decrease. If the fall in chromium pressure exercises a stronger influence on the surface concentration than the rise in arsenic pressure, then this will lead to a reduction in C_0.

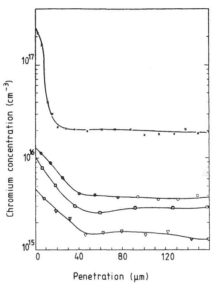

Figure 5.3 Chromium profiles, showing effect of adding extra arsenic to the ampoule. Diffusion temperature is 1000 °C and time is 4 h for all profiles: x, 0 μg; ○, 72 μg; □, 128 μg; ▽, 260 μg (from Tuck and Adegboyega 1979).

It is clear that development of the above argument requires a knowledge of the Cr–Ga–As phase diagram together with information on the various vapour pressures in equilibrium with the condensed phases. Progress in this direction was made by Deal *et al* (1985) in a piece of work which determined the ternary diagram from room temperature to 1300 °C for compositions of less than 50 atomic percent arsenic. The result for 800 °C is shown in figure 5.4. In this figure the widths of the one- and two-phase regions have been exaggerated for the sake of clarity. A feature of the diagram is the large number of solid phases which can exist. Deal *et al* pointed out that that GaAs is not in equilibrium with chromium at any temperature; to put it another way, there is never a tie-line joining the two phases. The most interesting feature from the point of view of diffusion work, however, is the invariant region, I, in which three phases co-exist, namely solid GaAs, solid CrAs and Cr–Ga–As liquid (see section 4.2.3). As noted in the previous chapter, the experimental profile should be the same for all systems described by points within the region.

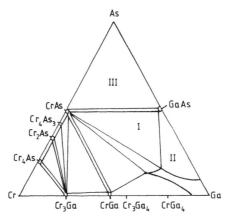

Figure 5.4   Ternary phase diagram for gallium–arsenic–chromium system at 800 °C (from Deal and Stevenson 1986).

Figure 5.4 goes a long way towards explaining some of the results of Tuck and Adegboyega (1979). Given the quantities of chromium and GaAs used in their experiments, the total system would have been described by a point in the invariant region. At the beginning of the heat treatment, time would have been required for the surface of the source and the surface of the GaAs to come into equilibrium with the ambient vapour. This presumably would have involved the formation of CrAs at the surface of the chromium and the creation of a small quantity of the liquid, probably also on the source surface, since both gallium and arsenic are more volatile than chromium. This process would proceed throughout the anneal so that, eventually, no metallic chromium remained. It seems likely that a quasi-equilibrium situation was approached quite quickly, at least as far as the ambient vapour pressures were concerned, since the surface concentration stabilised after a very short time, as noted above.

The data of figure 5.4 does not help explain the added-arsenic result of the earlier workers, however, since the phase diagram is only plotted for low percentages of arsenic. The upper half of the diagram must contain liquids at the temperatures of figures 5.1–5.3, since arsenic melts at 817 °C. It is certain that the addition of arsenic to the ampoule will take the system across the tie-lines joining CrAs and GaAs into this upper region (region III in the figure). To date, nothing is known about the phases present here or (more importantly) the ambient vapour pressures of chromium and arsenic in equilibrium with them.

Having established the major features of the Cr–Ga–As diagram, the same workers proceeded to use this information in a study of chromium diffusion in GaAs (Deal and Stevenson 1984, 1986). They took some pains over preparing diffusion sources within regions I and II in order to check that region I was indeed invariant. High resistivity undoped GaAs was used as the diffusion host and profiles were plotted using SIMS. A series of profiles from experiments using the same 'invariant' source is shown in figure 5.5. The experiments cover a wide range of diffusion times and only the first few microns of the profiles are plotted, in order to show the structure close to the surface. The general form of the curves is very similar to those of Tuck and Adegboyega; a surface peak drops sharply with distance from the surface to a horizontal bulk concentration. Between these two sections, a dip is observed in most of the profiles. This last feature was not observed by Tuck and Adegboyega, but has been seen in other systems, to be described later. The bulk concentration rises steadily with time and appears to be saturating at 24 hours (although care must be taken here to note that the chromium concentration scale in figure 5.5 is logarithmic, not linear). The values of C_0 and C_B are rather higher than those reported by the previous workers, and C_0 appears to vary with time in a non-systematic way, in contrast with the earlier work.

Experiments were also carried out using different sources taken from regions I and II. It was found that the bulk concentrations C_B were, as expected, the same for all sources within region I. The surface concentrations, C_0, showed a wide variation, however. A profile taken using a region II source showed much lower values of both C_0 and C_B. This latter result is consistent with figure 5.4, since the region II liquid contains less chromium and less arsenic than the region I liquid, so the vapour pressures of those two elements would be expected to be less for a region II diffusion. An important conclusion drawn from these experiments by Deal *et al* was that the value to which the bulk concentration approaches asymptotically represents the 'true solubility' of chromium in GaAs under the conditions of the experiment. This is an interesting proposal since, generally speaking, it is assumed that the surface concentration, C_0, gives this quantity. The surface peak is attributed by Deal *et al* to unspecified surface effects.

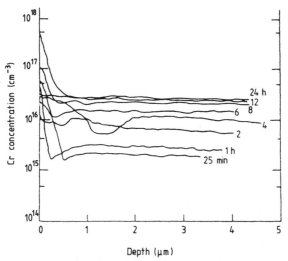

Figure 5.5 Diffusion profiles of chromium in GaAs for 800 °C (from Deal and Stevenson 1986).

Deeper SIMS profiles showed the bulk chromium levels rising with time, as in figure 5.5, but they also decreased deep in the crystal, as shown in figure 5.6, which gives a series of 'invariant source' profiles taken for different times at 850 °C.

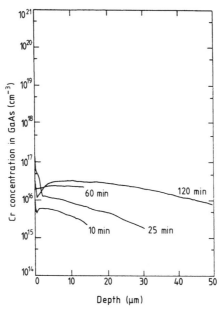

Figure 5.6 Diffusion profiles of chromium in GaAs for 850 °C, showing form of profile in bulk of crystal (from Deal and Stevenson 1986).

5.1.2 Out-diffusion studies

Experiments measuring the out-diffusion of chromium from doped GaAs have fallen into two main categories. In the first motivation has been to gather data on the diffusion process in order to develop a credible model. In the second the out-diffusion of chromium from a substrate into a growing epitaxial layer has been monitored. This latter type of experiment has ben pursued largely because of the effect such diffusion may have on devices subsequently produced in the epitaxial layer. In both cases, radio-tracer techniques and SIMS measurements have been utilised.

Tuck and Adegboyega (1979) chose 750 °C for their out-diffusion experiments, since this is a fairly typical processing temperature for the manufacture of field-effect transistors. The result of one such experiment is shown in figure 5.7. Initially a diffusion was performed at 1100 °C to create a chromium distribution within the specimen. The sample was then cleaved into three parts. The profile in the first piece was determined immediately and this is shown as curve A in the figure. The second part was annealed for one hour at 750 °C in an evacuated ampoule and then sectioned. The resulting profile is shown as curve B. Profile C was obtained after an anneal of four hours. The redistribution of chromium as a result of the anneals shows an interesting effect; not only does C_0 drop considerably, as might be expected, but the interior concentration also falls, indicating substantial transport of material from deep inside the crystal. It is particularly interesting to note that there is very little difference between the one-hour and the four-hour profiles.

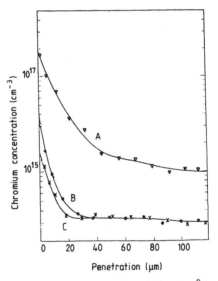

Figure 5.7   Out-diffusion of chromium at 750 °C. A is the original profile, B and C were taken after further anneals of 1 h and 4 h (from Tuck and Adegboyega 1979).

A second type of out-diffusion experiment was carried out using a homogeneous substrate. The specimen was prepared as follows. A radioactive diffusion was performed at 1100 °C, giving rise to a profile similar to those of figure 5.2. The highly doped surface regions on each side of the slice were then etched off, leaving a specimen homogeneously doped with radiotracer chromium to a level of 5×10^{16} cm^{-3}. The redistribution that took place after a one hour anneal is shown in figure 5.8. The interior concentration dropped considerably, but a new high concentration peak developed near the surface, fed by chromium diffusing rapidly from the interior of the crystal. It is important to note that the new surface concentration is greater than the original homogeneous doping level.

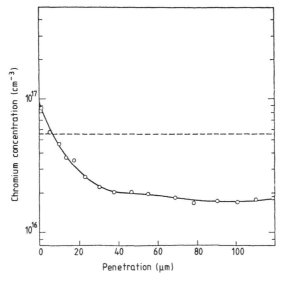

Figure 5.8 Out-diffusion from a sample which has been homogeneously doped with radiotracer chromium, anneal temperature 750 °C. The broken line shows the original chromium level (from Tuck and Adegboyega 1979).

The radiotracer approach was continued in a further paper by Tuck *et al* (1979) who again produced homogeneously doped samples by diffusion and used these as substrates for the epitaxial growth of thin layers of GaAs. Vapour phase epitaxy was employed as the growth method, using a conventional AsCl$_3$ system. The normal procedure for FET growth was adopted. Substrate temperatures were in the range 745–755 °C and a typical growth run, including cool-down, took about two hours. The level of chromium doping chosen for the substrate was in the region of 1 ppm, roughly equivalent to the concentration of chromium encountered in bulk-grown semi-insulating ingots. At the end of a growth run, the profile was obtained in the normal way, sectioning through the epitaxial layer and into the substrate. Results from two experiments in which 20 μm layers were grown are shown in figure 5.9; it can be seen that substantial quantities of chromium are found, even close to the surface.

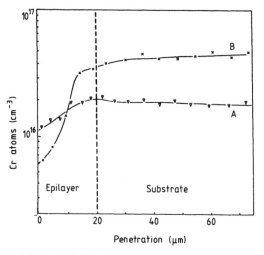

Figure 5.9   Out-diffusion during epitaxy of: A, undoped layers; and B, sulphur-doped layers (from Tuck *et al* 1979).

This result was subsequently checked using semi-insulating substrates which had been doped with chromium during growth rather than by diffusion. Linh *et al* (1980) grew MBE layers and used SIMS to detect chromium. With a substrate doping of about 1×10^{16} cm^{-3}, a chromium level of 2×10^{15} cm^{-3} was detected in the epitaxial layer. Also, as shown in figure 5.10, a substantial peak was found at the layer–substrate interface. Similar experiments were carried out using metal-organic chemical vapour deposition (MOCVD) to produce the epitaxial layers (Huber *et al* 1982). A similar out-diffusion effect was observed, although the interface peak was absent in this case. It is interesting to note that when the experiment was repeated using 'undoped' semi-insulating GaAs, the SIMS results still showed chromium in the substrate (at a level of 10^{15} cm^{-3}) and this still diffused out into the growing layer.

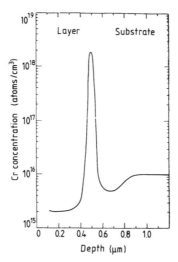

Figure 5.10 Chromium profile in MBE layer (from Linh *et al* 1980). Reproduced by permission of *Shiva Press*.

A number of workers have studied 'simple' out-diffusion from GaAs which has been doped with chromium during growth (Huber *et al* 1979, Vasudev *et al* 1980, Watanabe *et al* 1982, Wilson 1982, Deal and Stevenson 1986). For the most part, profiles were plotted using SIMS and surface peaks similar to that of figure 5.8 were found. Several researchers presented evidence to show that the out-diffusion result was dependent on whether the sample was capped with SiO$_2$ or free to exchange atoms with the ambient atmosphere (see, for example, Eu *et al* 1980). This would, of course, be expected in the light of the phase diagram discussion given above.

The most extensive of the investigations mentioned above is the one carried out by Deal and Stevenson (1986). Out-diffusions were performed in closed evacuated ampoules which also contained a large amount of 'sink' material. The composition of the sink was chosen to be such as to give rise to vapour pressures of zinc and arsenic which were insufficient to maintain the level of chromium doping within the GaAs slice. Thus GaAs with doping of 1×10^{17} cm^{-3} might be annealed with a sink of a composition corresponding to a point in region II of figure 5.4. Such a composition would be in equilibrium with a doping in the region of 5×10^{15} cm^{-3} in the semiconductor. This would be expected to give rise to out-diffusion from the slice, with a surface concentration level of 5×10^{15} cm^{-3} establishing itself quite quickly. The out-diffusion process might then be expected to proceed until this concentration is uniform throughout the slice.

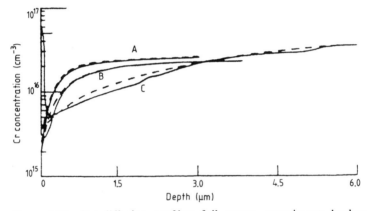

Figure 5.11 Out-diffusion profiles: full curves, experiment; broken curves, theory. A, 700 °C for 2 h; B, 800 °C for 1 h; C, 900 °C for 1 h (from Deal and Stevenson 1986).

Results for out-diffusions performed at 700 °C, 800 °C and 900 °C are shown in figure 5.11. Once again, a surface peak is seen to develop, with the new value of C_0 exceeding the original homogeneous concentration. Deal and Stevenson assumed that this peak was an artefact of the experiment and fitted error function curves to the remaining 'bulk' parts of the profiles. The erf curves are shown dotted in the diagram; the argument giving rise to these theoretical profiles will be given in section 5.1.4. In a few cases, reported by other workers, out-diffusion experiments have not shown the characteristic surface peak. Ugandawa *et al* (1980) annealed chromium-doped semi-insulating GaAs in the temperature range 700–900 °C. When the anneal was performed in an ambient of hydrogen, the form of the resulting profile depended on temperature. No redistribution was observed at 700 °C, while surface depletion was seen at 800 °C. Temperatures of 850 °C and 900 °C showed surface peaks. Experiments carried out in large arsenic overpressures showed no surface peaks. There is much evidence showing that when a slice is taken to high temperature at dissociation pressure of arsenic, the surface can lose a good deal of material (see, for instance, Tuck and Badawi 1979) and Ugandawa *et al* did, in fact, report the presence of gallium droplets on the surface at the end of this type of experiment. This observation perhaps explains the inconsistent nature of those particular results. It cannot explain, however, the results taken in an ambient of arsenic; in these tests, one would have expected the GaAs surface to be unaffected by the heat treatment. Ugandawa *et al* also used normal error functions to describe their profiles.

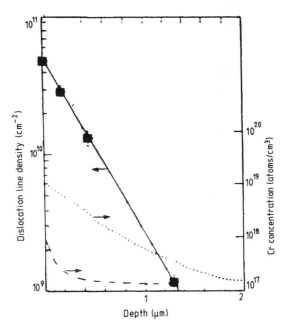

Figure 5.12 Defect density and chromium concentration profiles of back surface damaged annealed GaAs substrates (from Magee *et al* 1979). Broken curve, Cr (no anneal); dotted curve, Cr annealed at 750 °C for 1 h. Reproduced by permission of *Am. Inst. Phys.*

An interesting indication of the interaction between chromium out-diffusion and mechanical damage was given in papers by Magee *et al* (1979, 1980). Surface damage was induced on chromium-doped slices by mechanical polishing, which introduced dislocation networks extending about 1.5 μm below the surface. The slices were then subjected to heat treatments within the range 300–900 °C. Transmission electron microscopy (TEM) was used to study the dislocation networks and SIMS profiles were plotted so as to monitor movement of the chromium. A result showing the chromium profiles before and after annealing for one hour at 700 °C is shown in figure 5.12. Also shown in the figure is the dislocation density as a function of depth. It can be seen that an accumulation of chromium occurs throughout the damaged region. The effect was observed at temperatures as low as 300 °C. It is also interesting that Magee *et al* (1979) reported that the TEM examinations of the annealed samples showed no evidence of precipitation along dislocation lines or in discrete segregation zones. This is surprising in view of the very large chromium concentrations registered in figure 5.12. They interpreted this result as indicating that the chromium was present either in the form of complexes or in regions 'not readily detectable by conventional TEM techniques'.

5.1.3 *Electrical properties of diffused material*

The electrical action of chromium in as-grown GaAs is now understood fairly well; it forms a deep acceptor which can give rise to semi-insulating properties in the crystal. It does not follow, however, that atoms introduced by diffusion necessarily have the same effect. The point is of some significance to the semiconductor technologist, who wants to know how the (rather rapid) movement of chromium atoms during heat treatments is likely to affect device performance. A few reports have been published specifically studying electrical properties of diffused material.

Tuck and Adegboyega (1979) used a capacitance–voltage (C–V) technique to produce profiles of carriers which could be compared directly with atomic profiles. The mercury-probe method, mentioned in section 2.4, was used. In this technique the Schottky barrier is made by bringing the semiconductor into contact with a column of mercury. When the C–V measurements have been taken, it is a simple matter to unmake the barrier and remove the next layer from the slice by etching. A new set of measurements can then be taken by making contact with the new surface. Repetition of the process allows the carrier concentration profile to be produced. One such profile is shown in figure 5.13, following a diffusion for 15 minutes at 1100 °C into an n-type sample. The original n-type doping of 2×10^{16} cm^{-3} is also shown in the figure, together with the chromium atomic profile. Close to the surface, the resistivity of the slice is too great for carrier concentrations to be determined by the C–V method, while at greater depths the crystal is n-type, the electron concentration tending to the original 2×10^{16} cm^{-3} level. The high-resistance surface layers corresponds to the region in which the chromium concentration exceeds the original n-type doping.

A similar experiment was performed by Deal and Stevenson (1984), who diffused chromium into tellurium-doped GaAs of

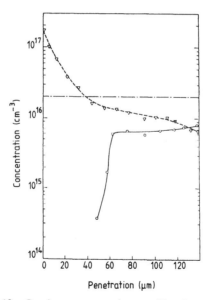

Figure 5.13 Carrier concentration profile after a 15 min diffusion with chromium. Diffusion temperature, 1100 °C. The corresponding atomic concentration profile is shown as the broken curve. Also shown is the original n-type doping level of 2×10^{16} cm^{-3} (from Tuck and Adegboyega 1979).

electron concentration 3.8×10^{17} cm^{-3}. After the diffusion anneal, 10 μm was etched from each surface to remove any surface peaks, leaving a sample homogeneously doped with chromium to a level which Deal and Stevenson estimated at 8×10^{16} cm^{-3}, the estimate being based on their previous experiments. A standard C–V measurement was then carried out on the sample, which was found still to be n-type, but with a reduced electron concentration of 2.4×10^{17} cm^{-3}. They pointed out that this result was consistent with the chromium acting as a double acceptor, as proposed by Brozel *et al* (1978).

Another group of experiments using a profiling method was reported by Ke *et al* (1985). Here the measurement technique employed was Deep Level Transient Spectroscopy (DLTS). This technique also involves a Schottky barrier, measuring the transient response of capacitance to a voltage pulse over a range of temperatures. The method gives information on the concentrations of deep levels in the semiconductor and also on the positions of the levels in the forbidden gap. Unfortunately, the mercury probe is not suitable for this type of measurement, so it was necessary to perform these experiments using 'permanent' Schottky contacts, made by evaporating gold onto the semiconductor surface. A DLTS measurement was then carried out, providing data on the deep levels just below the contact. The gold contact was then removed and a layer was etched from the GaAs before evaporating another. In this way, a profile was produced, giving concentration of deep levels as a function of depth.

Diffusions were carried out using radiotracer chromium. At the end of a diffusion, the specimen was divided, one half being used for radiotracer evaluation and the other for the DLTS measurements, again permitting comparison to be made between atomic and electronic profiles. Two main deep levels were found in this work: an electron trap, usually called EL 1, and a hole trap. Good agreement between atomic concentration and that of the hole trap was found for diffusions carried out at 1000 °C, as can be seen in figure 5.14. Ke *et al* (1985) took this result to mean that most of the chromium atoms were incorporated in the GaAs crystal as a single acceptor species. For specimens diffused at lower temperatures, they found that only a fraction of the chromium introduced deep acceptor levels.

Indirect evidence of the acceptor action of diffused chromium is provided by photoluminescence experiments (Skolnick *et al* 1982, Fujiwara *et al* 1985). The 4 K photoluminescence spectrum of GaAs doped in growth with chromium is dominated by a strong line at 0.84 eV together with associated phonon side-bands at lower energy. Measurements on material diffused at 1000 °C showed a strong photoluminescent response, with the intensity of the 0.84 eV line roughly proportional to the concentration of chromium atoms. It is interesting, however, that Skolnick *et al* reported that the 0.84 eV line was only observed in samples diffused at 800 °C and above. They suggested that for low temperature diffusions most of the chromium atoms occupy electrically inactive sites. This is not the only conclusion that can be drawn from their result, however. It is generally believed that the centre giving rise to the luminescent line is not simply chromium on a gallium site, Cr_{Ga}, but a complex involving Cr_{Ga} and an arsenic vacancy. It may well be, therefore, that after a low temperature diffusion all of the chromium is on gallium sites but the complex has not been able to form.

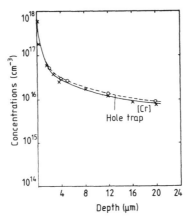

Figure 5.14 Profiles of chromium concentration and the chromium-associated hole trap concentration after a 1 h diffusion at 1000 °C (from Ke *et al* 1985).

5.1.4 *Modelling of the diffusion mechanism*

Most workers have assumed that chromium diffuses in GaAs using some version of the substitutional–interstitial mechanism (see section 4.1). Two main approaches have been employed. The first is a direct computer modelling technique, similar to that outlined in section 4.2.2; this is so similar to that proposed for silver in the III–Vs, however, that discussion of this particular model will be postponed until Chapter 6. The second approach is by way of Fick's law. As noted in the previous chapter, this can lead to some rather intractable equations, but progress can be made by assuming that simplifications can be made for specific regions of the crystal, different simplifications being used for different regions. This technique has been tried by Tuck and Adegboyega (1979) and by Deal and Stevenson (1986), and the version given by the latter group will be presented here. The method owes a good deal to an earlier piece of theory by Sturge (1959), originally proposed to describe the diffusion of copper in germanium.

The assumptions are similar to those for diffusion of zinc in GaAs, outlined in the previous chapter. Most of the chromium atoms occur on the gallium site, Cr_{Ga}, giving rise to a species which has negligible diffusivity. The minor species is an interstitial, Cr_i, which can move through the lattice very quickly. Interchange between the two species involves gallium vacancies, as follows:

$$Cr_{Ga} \rightleftharpoons Cr_i + V_{Ga}. \tag{5.1}$$

In contrast with the earlier case, it is assumed that neither of the species is charged. Equation (5.1) gives rise to the mass-action relation

$$KS = IV \tag{5.2}$$

where S, I, are the concentrations of substitutional and interstitial chromium atoms, V is the concentration of gallium vacancies, and K is the equilibrium constant (see equation 4.3). For the constant surface concentration case, the quantities S, I, V, are all fixed at the surface, which is designated $x = 0$. If these surface values are indicated by a prime, we have an expression for K:

$$K = \frac{I' V}{S'}. \tag{5.3}$$

All transport of material is due to interstitial diffusion, so equation (4.6) still applies

$$\frac{\partial I}{\partial t} + \frac{\partial S}{\partial t} = D_i \frac{\partial^2 I}{\partial x^2} \tag{5.4}$$

where D_i is the interstitial diffusion coefficient. Similarly the continuity equation for vacancies, equation (4.11), is also operative

$$\frac{\partial V}{\partial t} = D_v \frac{\partial^2 V}{\partial x^2} - \frac{\partial S}{\partial t} + k(V' - V). \tag{5.5}$$

D_v is the diffusion coefficient for vacancies and k is a constant giving the rate of production of vacancies in the bulk of the crystal.

As noted earlier, the simultaneous solution of these equations is difficult and it proves to be helpful to consider a number of special cases.

(a)   Vacancy equilibrium is maintained. The production of vacancies in the bulk is then so efficient that the crystal always has the correct number, i.e. $V = V'$ for all x. Equation (5.2) then becomes

$$I = \frac{KS}{V'} \tag{5.6}$$

which, substituted into equation (5.4) gives

$$\frac{\partial S}{\partial t} \left(1 + \frac{V'}{K} \right) = D_i \frac{\partial^2 S}{\partial x^2}. \tag{5.7}$$

It has already been assumed that only a small fraction of the diffusing element is in the interstitial form at any one time. It follows that S is approximately equal to the total impurity concentration and equation (5.7) can be taken as the diffusion equation for the chromium atoms.

Rearranging equation (5.7) and substituting for K from equation (5.3),

$$\frac{\partial S}{\partial t} = D_i \left(\frac{I'}{I' + S'} \right) \frac{\partial^2 S}{\partial x^2} \tag{5.8}$$

giving an effective diffusion coefficient

$$D_1 = D_i \left(\frac{I'}{I' + S'} \right) \tag{5.9}$$

(b)   No production of vacancies in the bulk, i.e. $k = 0$. This is the other extreme. Here the vacancies needed in the reaction given by equation (5.1) are required to diffuse in from the surface. The mechanism becomes one in which a substitutional atom dissociates into an interstitial and a vacancy which diffuse independently. Since, by hypothesis, vacancy diffusion is a slow process compared to that of interstitials, it can be assumed that the interstitial concentration becomes I' throughout the whole crystal. This amounts to assuming an infinite interstitial diffusion coefficient. Putting $I = I'$ in equation (5.2) and substituting for V in equation (5.5), with $k = 0$, we obtain

$$\frac{\partial S}{\partial t}\left(1 + \frac{I'}{K}\right) = D_v \frac{\partial^2 S}{\partial x^2} \tag{5.10}$$

which, on substituting for K, gives an effective diffusion coefficient

$$D_2 = D_v\left(\frac{V'}{V' + S'}\right) \tag{5.11}$$

i.e. the over-all diffusion process is determined by the diffusion coefficient of vacancies.

(c) $k > 0$, but still small. Here the assumption is retained that the interstitial coefficient is so much faster than any other that interstitial equilibrium establishes itself immediately, i.e. $I = I'$ for all x and t. Equations (5.2) and (5.5) then give

$$\frac{\partial S}{\partial t} = D_2 \frac{\partial^2 S}{\partial x^2} + \frac{S' - S}{\theta} \tag{5.12}$$

where

$$\theta = \frac{V' + S'}{kV'}.$$

Note that θ has the dimensions of k^{-1}, i.e. of time. For constant surface concentration conditions and a semi-infinite crystal, the boundary condition is given by

$$S = S' \text{ at } x = 0 \text{ for all } t.$$

Equation (5.12) can be solved for this condition using Laplace Transforms. The solution is

$$\frac{S}{S'} = 1 - \exp\left(-\frac{t}{\theta}\right) \text{erf} \frac{x}{2(D_2 t)^{1/2}} \tag{5.13}$$

(Note that the solution given by Deal and Stevenson is incorrect. Fortunately this does not affect any of the physical arguments given in their paper, since they did not believe that case (c) was operating in their experiments.)

(d) Solutions for large k (small θ). The crystal is able to produce vacancies in the bulk quite efficiently and can thereby help the substitutional–interstitial transition. It can no longer be assumed that the interstitial concentration acquires its equilibrium value I' instantaneously. The vacancy concentration V is still less than V', but the decrease is less than for case (c). We assume that the internal mechanism is so effective at replacing vacancies that diffusion from the surface can be ignored. A quasi-equilibrium can then be attained in which interstitial atoms join the lattice at the rate at which the mechanism creates vacancies, i.e.

$$\frac{\partial S}{\partial t} = k(V' - V). \tag{5.14}$$

Now consider the interstitial distribution. If k were so large that the vacancy concentration was always at its equilibrium value, V', we would have for the semi-infinite specimen

$$\frac{I}{I'} = \frac{S}{S'} = \text{erfc } \mu \tag{5.15}$$

where

$$\mu = \frac{x}{2(D_1 t)^{1/2}}.$$

If k is slightly smaller than this, it can be shown that, to a first approximation, I/I' is unchanged. Physically this amounts to saying that a lack of vacancies will interrupt the flow of atoms from the interstitial to the substitutional distribution, but will not interfere very much with the original interstitial distribution, providing the deviation from equilibrium is not too great.

Substituting $I = I'$ erfc μ into equation (5.2) and combining with equation (5.14) gives

$$\frac{\partial S}{\partial t} = \frac{kK}{I'}\left(S' - \frac{S}{\text{erfc } \mu}\right). \tag{5.16}$$

Substituting for k from equation (5.3) and approximating by

$$\theta \simeq \frac{S'}{kV'}$$

gives

$$\theta \frac{\partial x}{\partial t} = S' - \frac{S}{\text{erfc } \mu}. \tag{5.17}$$

The problem has now been simplified considerably, but further simplification is required to solve equation (5.17). Sturge (1959) treated μ as a constant and obtained the approximate solution.

$$S = S' \text{ erfc } \mu \left[1 - \exp\left(\frac{-t}{\theta \text{ erfc } \mu} \right) \right]. \qquad (5.18)$$

According to Sturge, the error involved in the approximation is small except in the region $t/\theta \simeq 10 \text{ erfc } \mu$. This solution must be used with some caution, since so many approximations were made in the course of its derivation.

Deal and Stevenson (1986) proposed that the conditions for chromium diffusion in GaAs in the bulk of the crystal (but not close to the surface) correspond most closely to case (d), above, and fitted their in-diffusion results to equation (5.18). In view of this proposal, it is worth looking a little more closely at the equation to determine whether it has a simple physical interpretation. The first term on the right-hand side represents the 'equilibrium' solution to the problem, i.e. the form of the profile obtained when vacancy equilibrium is maintained and $V = V'$ throughout the experiment. This term is modified by the expression in brackets, which is always less than unity. For all values of x and t, therefore, concentrations are less than the 'equilibrium' concentrations. This deficiency becomes less as time progresses. In particular, at the surface $(x = 0)$, the concentration is less than S', but approaches it with increasing time of diffusion. This qualitative behaviour is shown by the profiles of Deal and Stevenson if the bulk portion is extrapolated to the surface. Reasonable correspondence between the experimental 'bulk' profiles and equation (5.18) could be found by choosing suitable values for D_1 and θ. It should be noted, however, that the portion of experimental profile which was fitted by equation (5.18) always covered less than one order of magnitude in concentration. This falls far short of the rule of thumb given in section 2.9 for claiming good correspondence between experiment and theory.

For the region just below the surface, it was suggested that case (b) should apply. This would predict the equilibrium concentration at the surface, S', for all diffusion times and a shallow erfc profile meeting up with the 'bulk' profile of equation (5.18) just below the surface. Such behaviour was not observed by Deal and Stevenson and this they attributed to what they called 'unusual surface phenomena' obscuring diffusion in the surface region.

For their out-diffusion experiments, they suggested that case (b) conditions applied, so that the profiles should have been of the form (see equation (2.25))

$$S - S_1 = (S_2 - S_1) \operatorname{erfc} \frac{x}{2(D_1 t)^{1/2}} \qquad (5.19)$$

where S_1 is the original homogeneous doping of chromium and S_2 is the new surface concentration during the out-diffusion (note that $S_2 > S_1$ in this case). The experimental and theoretical out-diffusion profiles are shown in figure 5.11. It can be seen once again that good correspondence is found providing those parts of the profile close to the surface are ignored.

An alternative explanation for the surface pile-up which occurs on annealing was presented by Yee *et al* (1983). They pointed out that if a diffusing species is charged, its equilibrium distribution near the surface will be affected by any electric field due to surface states. They developed this idea assuming that the diffusing chromium exists as either positive or negative ions and showed that it could produce the observed surface peak. The theory requires the penetration of the peak into the specimen to be of the same order as the Debye length, i.e. rather less than 1 μm, and this prediction agrees well with the reported SIMS work. It should be noted, however, that in the radiotracer work of Tuck and Adegboyega (1979) much deeper peaks were observed.

5.1.5 *Chromium in GaAs: concluding remarks*

Because of its significance in device manufacture, this system has been the subject of a very large number of published papers. Unfortunately, the great majority of these have reported a rather restricted range of results and there have been few large-scale investigations. Some findings have been confirmed many times, however, so we can take them to be correct. Foremost among these is the remarkable inclination shown by chromium to pile up at interfaces. It will be shown later in this book that this inclination is shared by other fast-diffusing elements. Pile-up occurs at sample surfaces, interfaces between substrates and epitaxial layers and, especially, in damaged regions. This last result strongly suggests that the enhanced solubility is associated with the presence of defects. The phenomenon is observed both during in-diffusion and out-diffusion from previously doped substrates and has been demonstrated using both radiotracer and SIMS techniques. It is interesting that the penetration of surface peaks determined by the first of these methods seems to be much greater than that found by the second. It is not clear whether this is because slightly different experiments were carried out in the first place or because of artefacts of the techniques. In any case, it is certainly significant that the same qualitative result has been obtained from two entirely different methods.

Many authors seem to assume that the high concentrations measured in the interface peaks are spurious in some way, and take the bulk concentration to be the 'true' solubility of chromium in GaAs under the conditions of the experiment. One or two of the results quoted here give rise to doubts concerning this conclusion. There is evidence, for instance, that a high proportion of the chromium in a peak can be electrically active, implying that it is occupying the normal site in the crystal, although there is also an indication that the chromium is much less active if diffused at low temperature. The failure to observe precipitation also seems to imply that the chromium may still be dissolved in the lattice even at high concentration. All this has a great bearing on the modelling of the diffusion process, of course, and on the validity of the models which have been proposed. The problem is far from being resolved and requires further work.

5.2 Manganese in GaAs

5.2.1 Early work

Vieland (1961) studied p–n junction depths for various acceptors and donors diffused into GaAs under arsenic pressures ranging from 0.1 atm to 10 atm. In general, the junction depths for manganese were found to decrease with increasing arsenic pressure, with a plateau occurring at the lowest pressures used. Radiotracer profiles were plotted by Seltzer (1965) over the temperature range 850–1100 °C for 'infinite source' conditions. He reported two types of profiles. One type fitted an erfc curve quite well and an activation energy of 2.49 eV was quoted for this category of profile. The other had a concave section, very like the shape found for zinc profiles, described in Chapter 4; examples are shown in figure 5.15 for 900 °C diffusions. This latter type was analysed by fitting two erfc curves to each profile and defining a 'fast' and a 'slow' component of diffusion, each giving rise to its own erfc curve. The two diffusivities differed by a factor of about five at 900°C and both were found to decrease with increasing arsenic pressure, in agreement with the work of Vieland.

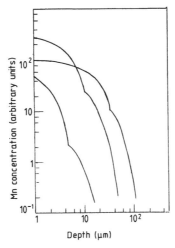

Figure 5.15 Diffusion profiles of Mn in GaAs at 900 °C, each one fitted to a pair of ERFC curves (from Seltzer 1965). Reproduced by permission of *Pergamon Journals Ltd*.

The procedure of fitting an experimental diffusion profile with more than one erfc has been adopted quite frequently in the literature. Usually it is proposed that each erfc is due to a different diffusing species. In general, such a result would be obtained only if the various species diffused without interacting; in most cases this must be considered rather unlikely. Seltzer proposed a mechanism in which diffusion of manganese takes place by the complex $(V_{As}Mn_{Ga}V_{As})^+$. He believed that both his 'fast' and 'slow' profiles were due to this mechanism, although he did not understand why it should give rise to two different values of diffusion coefficient. In a later paper, Kendall (1968) suggested that Seltzer's results could be explained satisfactorily by a substitutional–interstitial mechanism and subsequent workers have tended to support this view.

5.2.2 Type conversion

When GaAs is annealed at temperatures in excess of about 700 °C, changes in conductivity are likely to occur. The phenomenon comes into that irritating category of effects which are not quite reproducible. GaAs samples from different sources show different conversion characteristics as, indeed, do samples obtained from the same source at different times, and sometimes samples can be raised to quite high temperatures without major changes in conductivity. Most early reports of type conversion concerned changes of n-type material to p-type after heat treatment (Edmond 1960, Hwang 1968, Toyama 1969). At that time it was commonly suggested that copper was the culprit, diffusing into the semiconductor from some source within the annealing furnace. More recently, interest has shifted to semi-insulating material, either chromium-doped or undoped. The most frequent report has been of conversion to p-type conductivity, but conversion to n-type has also been found on occasion (Tuck and Adegboyega 1979, Mircea-Roussel *et al* 1980) and, in one report, it was found that an initially semi-insulating slice could have a thin n-type skin at the surface with p-type conversion below (Ohno *et al* 1979). An interesting discussion has taken place in the literature concerning the role of manganese in the phenomenon and most of the consequent experimental work has been pursued in an attempt to throw some light on this role.

It is generally agreed that manganese normally occupies the gallium site in GaAs and acts as an acceptor, giving rise to an energy level about 0.1 ev above the valence band. It therefore turns GaAs p-type if it is the dominant dopant and, in addition, it provides a characteristic photoluminescence spectrum, showing a principal line of 1.41 eV at 4 K, together with associated phonon replicas. A number of groups reported an apparent correlation of p-type conversion at the surface of semi-insulating GaAs and the appearance of a photoluminescent signal identical with the 1.41 eV manganese line (Hallais *et al* 1977, Zucca 1977, Mircea-Roussel *et al* 1980). The reasonable conclusion was drawn that the conversion was due to manganese acceptors at the surface of the specimen, although initially it was not clear whether the element appeared because of diffusion from outside or by out-diffusion of manganese present as an accidental dopant in the original crystal.

Klein *et al* (1980) carried out a series of experiments in which undoped semi-insulating GaAs was subjected to a variety of heat treatments. Manganese was detected in the slices using both photoluminescence and SIMS. Surprisingly, the conversion effects which were observed proved to depend on the ambient atmosphere. Material heated in hydrogen at 740 °C turned p-type at the surface, while slices heated in argon remained high resistivity n-type. A SIMS profile taken after heat treatment in hydrogen for 90 minutes at 740 °C is shown in figure 5.16. The profile from a sample heated in argon, on the other hand, was similar to that for an unheated slice. Figure 5.16 resembles the chromium profiles of figures 5.2 and 5.8, but it is not clear from its form whether it is due to in-diffusion or out-diffusion. Polishing a few microns from the surface removed the manganese and restored the n-type high resistivity; it also removed the 1.41 eV luminescence. A further heat treatment turned the surface p-type again.

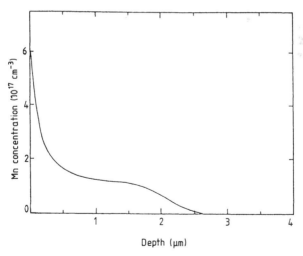

Figure 5.16 SIMS Mn profile in undoped semi-insulating GaAs after anneal in hydrogen at 740 °C (from Klein *et al* 1980).

If the origin of the manganese was outside the slice, the most likely sources would be either the hydrogen ambient or the furnace lining. Separate experiments were therefore performed in vacuum and also in hydrogen using a PBN liner shielding the sample from the quartz furnace tube. Both types of experiment gave rise to conversion at the surface and a manganese profile similar to figure 5.16. Klein *et al* came to the conclusion that the manganese came from within the sample. Figure 5.16 was therefore presented as an out-diffusion profile, the element occurring because of manganese impurity in the materials used in the original crystal growth. The resemblance between the manganese profiles and those found by other workers for out-diffusion of chromium from GaAs suggested a similar diffusion mechanism for the two elements.

5.2.3   A model for manganese out-diffusion

An attempt to model the experimental results of Klein *et al* (1980) has been made by Jordan (1982), using yet another variation of the substitutional–interstitial mechanism. Jordan reduces the complexity of the analysis a little by assuming all species to be neutral. (The same simplification was made by Deal and Stevenson 1986.) The basic defect interaction between the two types of atom is therefore

$$V_{Ga} + Mn_i \rightleftharpoons Mn_{Ga}. \tag{5.20}$$

In sections 4.1 and 5.1.4 it has been assumed that the quasi-chemical reaction is always in equilibrium so the law of mass-action could be used. Jordan rather heroically dropped this simplification, thereby making the analysis much more difficult. He assumed that the equation proceeds right–left with a rate k_1 and left–right at rate k_2. The continuity equations for substitutional and interstitial atoms become

$$\frac{\partial S}{\partial t} = k_2 VI - k_1 S \tag{5.21}$$

$$\frac{\partial I}{\partial t} + k_2 VI - k_1 S = D_i \frac{\partial^2 I}{\partial x^2} \tag{5.22}$$

where S and I are concentrations of manganese substitutionals and interstitials, and V is the gallium vacancy concentration (note that adding together equations (5.21) and (5.22) gives equation (5.4)).

Jordan assumed a negligible creation rate for vacancies, so the equation for vacancies is

$$\frac{\partial V}{\partial t} + k_2 VI - k_1 S = D_v \frac{\partial^2 V}{\partial x^2}.$$ (5.23)

All that is required now is to provide a simultaneous solution to equations (5.21)–(5.23) subject, of course, to the appropriate boundary and initial conditions. This proves to be rather too difficult and, having started very rigorously, the simplifications and approximations now come thick and fast in order to produce analytical expressions.

Consider the out-diffusion of manganese from a slice of thickness $2l$, with surfaces at $x = \pm l$. The sample is initially homogeneous and the assumption is made that the concentrations of manganese interstitials and of gallium vacancies are zero so that all the manganese in the sample is in the substitutional state, i.e.

$$\left.\begin{array}{l} I = 0 \\ V = 0 \\ S = S^0,\ \text{say} \end{array}\right\} \text{ for all } x.$$

Jordan used this set of starting conditions in the belief that they are a reasonable representation of the room-temperature state of a crystal. At time $t = 0$ the sample is raised to diffusion temperature and the gallium vacancy concentration at the surface goes instantaneously to the thermal equilibrium value V', i.e.

$$\begin{array}{llll} \text{at } t = 0 & I = 0 & \text{all } x \\ & S = S^0 & \text{all } x \\ & V = V' & \text{at } x = \pm l \\ & V = 0 & \text{elsewhere.} \end{array}$$

Jordan next draws our attention to the very different conditions existing in the centre and at the surfaces of the slice. In the bulk, the initial conditions are such as to drive equation (5.20) to the left, i.e. to make atoms leave the lattice and go interstitial. Equation (5.21) becomes

$$\frac{\partial S}{\partial t} = - k_1 S$$ (5.24)

which has solution

$$S = S^0 \exp(- k_1 t)$$ (5.25)

and, since the concentrations of interstitial and substitutional atoms must initially sum to S^0, the solution for interstitials is

$$I = S^0[1 - \exp(-k_1 t)]. \qquad (5.26)$$

It can be seen that, strictly, equations (5.25) and (5.26) are correct only for $t \approx 0$. Outside this range they are approximations which deteriorate as time progresses.

It is assumed that no atoms leave the surfaces, so that the quantity of manganese inside the slice is constant throughout the diffusion. This gives the boundary condition

$$\frac{\partial I}{\partial x} = 0 \quad \text{at } x = \pm l. \qquad (5.27)$$

It is also assumed that close to the surface equation (5.20) moves to the right due to the high concentration of vacancies close to $x = \pm l$ eating up any available interstitials. The concentration of substitutionals close to the surface can therefore be found by integrating equation (5.21)

$$S = k_2 \int_0^t IV \, dt + S^0. \qquad (5.28)$$

Again, this is strictly correct only for small values of t. If expressions can be found for I and V as functions of x and t, the concentration of substitutional manganese close to the surface can be found for small diffusion times.

If we take equation (5.26) as representing the concentration of interstitials at the centre of the slice, use the boundary condition (5.27) and also ignore any interchange between interstitials and vacancies, the interstitial distribution can be determined from Fick's law:

$$\frac{\partial I}{\partial t} = D_i \frac{\partial^2 I}{\partial x^2}. \qquad (5.29)$$

Jordan gives the following solution

$$I = S^0(1 - e^{-k_1 t}) + S^0 \frac{4}{\pi} \sum_{n=0}^{\infty} \frac{(-1)^n}{2n+1} \cos \frac{(2n+1)\pi(l-x)}{2l}$$

$$\times \left[\frac{k_1 \tau}{(2n+1)^2} + \left(\frac{k_1 \tau}{(2n+1)^2} \right)^2 + \cdots \right] \left[e^{-(t/\tau)(2n+1)^2} - e^{-k_1 t} \right] \qquad (5.30)$$

where

$$\tau = \frac{4l^2}{D_i \pi^2}.$$

Even this rather impressive expression must be taken as a fairly rough approximation, however, because of the simplifications which have been made in the derivation.

The vacancy distribution is determined by solving equation (5.23) using a Laplace Transform technique. The simplification made here is to assume that the interstitial concentration can be taken as essentially constant for the purposes of the calculation. (There is some internal consistency here; equation (5.30) does, in fact, give a very flat profile near the surface.) The result given by Jordan is:

$$V = \frac{V'}{2} \sum_{n=0}^{\infty} \left[\exp\{ - [(2n+1)l - x]B^{1/2}\} \operatorname{erfc}\left[\frac{(2n+1)l - x}{2(D_v t)^{1/2}} - (D_v Bt)^{1/2} \right] \right.$$

$$\left. + \exp\{ [(2n+1)l - x]B^{1/2}\} \operatorname{erfc}\left[\frac{(2n+1)l - x}{2(D_v t)^{1/2}} + (D_v Bt)^{1/2} \right] \right]$$

$$+ \frac{V'}{2} \sum_{n=0}^{\infty} \text{ (expression as above, with} - x \text{ replaced by} + x) \quad (5.31)$$

where

$$B = \frac{k_2 I}{D_v}.$$

In principle, equations (5.28), (5.30) and (5.31) can be used to calculate the manganese out-diffusion profile in GaAs. This calculation was performed using a numerical procedure and the result was compared with the experimental data of Klein *et al* (1980). Because of the various approximations made, the solution was valid only close to the surface. It was assumed that the concentration remained constant in the bulk region; the cut-off depth at which the 'surface' region gives way to the 'bulk' region was decided somewhat arbitrarily by choosing the depth at which the integral in equation (5.28) falls below 10 percent of S^0. The constant 'bulk' level was easily determined from the requirement that the quantity of manganese within the slice remains the same throughout the diffusion.

The final result is shown in figure 5.17, together with the experimental points of Klein *et al*. The constants used in the calculation are: $T = 740\,^{\circ}\text{C}$, $D_i = 10^{-6}\,\text{cm}^2\,\text{s}^{-1}$, $D_v = 8 \times 10^{-10}\,\text{cm}^2\,\text{s}^{-1}$, $k_1 = 10^{-5}\,\text{s}^{-1}$, $k_2 = 2.8 \times 10^{-14}\,\text{s}^{-1}\,\text{cm}^3$, $S^0 = 2 \times 10^{15}\,\text{cm}^{-3}$, and $V' = 10^{14}\,\text{cm}^{-3}$. The agreement between experiment and theory is reasonable down to a depth of about 0.5 μm. The apparent discontinuities which occur at the interface between the surface and bulk regions are a consequence of the cut-off procedure.

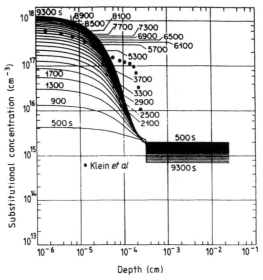

Figure 5.17 Calculated Mn profile in annealed GaAs slice, with an initial background level of $2 \times 10^{15}\,\text{cm}^{-3}$. The curve of figure 5.16 (5400 s) is shown for comparison (from Jordan 1982). Reproduced by permission of *Shiva Press*.

5.2.4 Diffusion in heavily doped material

Surprisingly, experiments in which manganese was diffused into heavily doped GaAs proved to give rather more simple results than either the work of Seltzer (1965) or Klein *et al* (1980), both of which were, on the face of it, more straight-forward approaches to the problem. Skoryatina (1986) diffused manganese into material which had been doped either p-type with zinc or n-type with tin

during crystal growth. Experiments were carried out in the temperature range 950–1100 °C and the diffusion source was a solution
of radioactive manganese chloride deposited on the surfaces of the
samples. Radiotracer profiles were plotted covering two orders of
magnitude in concentration and erfc curves were reported for all
cases, so that values of diffusion coefficient could be unambiguously assigned. The variation of diffusion coefficient with temperature is shown in figure 5.18 for three types of original impurity
content: zinc doping of 3.4×10^{19} cm^{-3} (line A) and 1×10^{19} cm^{-3}
(line B) and tin doping of 5×10^{17} cm^{-3} (line C). All three followed
the relation

$$D = D_0 \exp\left(\frac{-Q}{kT}\right) \tag{5.32}$$

with values of D_0 of 7.3×10^{-5}, 3.2×10^{-4} and 3.9×10^{-2} cm^2 s^{-1}
respectively and activation energies of 1.24 eV for both p-type
crystals and 2.25 eV for the n-type. It should be noted that the ratio
of the diffusion coefficients in the p-type samples is the same over
the whole temperature range and that this ratio corresponds closely

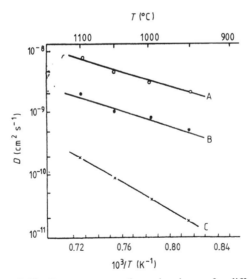

Figure 5.18 Temperature dependencies of diffusion
coefficient of Mn in GaAs doped with zinc (A, B) and with
tin (C) (from Skoryatini 1986). Reproduced by permission
of *Am. Inst. Phys.*

to that of the original p-type dopings, i.e. a factor close to 3.4. This strongly suggests a diffusion coefficient proportional to hole concentration.

For the p-type samples, an explanation was suggested that was essentially the substitutional–interstitial model of section 4.1. The experiments were virtually isoconcentration (see section 2.5.2), since the original p-type doping was sufficient to dominate the electrical character of the samples even at 1100 °C. The only difference between these experiments and the 'genuine' zinc isoconcentration tests described in section 4.2.1 is that the acceptor manganese was used instead of the acceptor zinc in the isoconcentration diffusion. The essence of both experiments, however, is that the position of the Fermi level in the forbidden gap is fixed throughout by the zinc content.

It was proposed that the manganese is electrically neutral in the interstitial form and singly ionised on the lattice, so the equivalent form of equation (4.1) becomes

$$s^- + h \rightleftharpoons i + v. \qquad (5.33)$$

Applying the same line of argument to equation (5.33) as that previously applied to equation (4.1), and remembering that the hole concentration p is constant in these experiments, gives an expression for effective diffusion coefficient (cf. equation 4.9)

$$D = \frac{K_1 D_i p}{V'} \qquad (5.34)$$

i.e. a coefficient proportional to hole concentration in agreement with experiment.

The diffusion behaviour of the n-type samples is less straightforward, but quite consistent with the above explanation. The complication is that the doping used in this case is not very different from the intrinsic carrier concentration at diffusion temperature. Thus at 900 °C the carrier concentration probably was dominated by the impurity doping and the ratio of diffusion coefficients at this temperature is roughly according to equation (5.34). At 1100 °C, on the other hand, the sample would have been intrinsic, with a hole concentration rather greater than $3 \times 10^{17} \, \text{cm}^{-3}$. The ratio recorded is therefore less than at the lower temperature.

It is, perhaps, a little surprising that these diffusions look quite so much like real isoconcentration experiments, with none of the signs of defect breakdown indicated by the work of of Seltzer (1965). One possible explanation is that Seltzer's profiles covered a rather greater range of manganese concentration; it is a sad fact of the experimenter's life that the more sensitive his experiments, the more difficult it is to fit his results to well-known theories.

5.3 Iron in GaAs

Several groups of workers have used radiotracer techniques to study diffusion of this element (Uskov and Sorvina 1974, Boltaks *et al* 1975, Brozel *et al* 1981, 1983). They all reported a high rate of diffusion, with profiles resembling either Zn/GaAs or Cr/GaAs, strongly suggesting the substitutional–interstitial mechanism at work again. Examples of profiles taken at 900 °C and 1000 °C are shown in figure 5.19. Boltaks *et al* found that the distribution in the bulk of the specimens (i.e. ignoring the profile close to the surface) could be fitted reasonably well to erfc curves and were able to

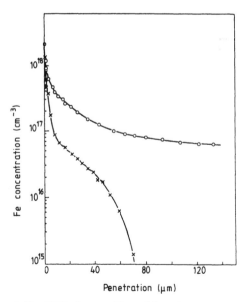

Figure 5.19 Diffusion profiles of iron in GaAs after 1 h: x, 900 °C; ○, 1000 °C (from Brozel *et al* 1983).

derive the Arrhenius expression:

$$D = 4.2 \times 10^{-2} \exp\left(\frac{-1.8 \pm 0.1 \text{ (eV)}}{kT}\right) \text{cm}^2\text{s}^{-1}. \quad (5.35)$$

It has been noted several times already that fitting a standard solution to only a part of a profile is a somewhat dubious procedure. It does, at least, give some idea of the penetration of iron obtained in their experiments, however, and to that extent it can be helpful.

Iron that has been added to a GaAs melt before commencement of crystal growth is known to act as a deep acceptor with an associated energy level at about 0.52 eV above the valence band edge (Allen 1968). It is generally assumed that the iron is accommodated on the gallium site and if used to compensate shallow donors, it acts in much the same way as chromium, producing high-resistivity material. Brozel *et al* (1983) carried out a series of experiments to determine whether atoms introduced by diffusion behaved in the same way. Electron paramagnetic resonance revealed a high concentration of Fe_{Ga} in the diffused specimens, strongly suggesting that the element had entered the lattice in the normal manner. Electrical measurements were also made on the diffused specimens using a C–V method. The results were rather similar to those shown in figure 5.13 for chromium-doped material; for iron concentrations in excess of the original n-type doping, the material became high-resistivity. As the iron concentration dropped below the original doping deep inside the crystal, the sample reverted to n-type, with the electron concentration corresponding to the level prior to diffusion. A detailed analysis of the carrier removal, shown in figure 5.20, gives an estimate of the fraction of iron atoms acting as acceptors. The figure suggests that for iron concentrations in excess of the original n-type doping, electrons. For the higher concentrations (corresponding to the portion of diffusion profile close to the surface) the 'active fraction' is much less than 0.5, however. Brozel *et al* were unable to account for this relatively low activity and suggested that where the iron is in high concentration it might be involved in a complex of some kind.

Out-diffusion experiments have been carried out by Huber *et al* (1982), using SIMS. Epitaxial GaAs layers were grown by MOCVD on GaAs substrates which contained iron. Out-diffusion of iron into the layer was demonstrated and profiles looking very much like that of figure 5.10 were produced, showing a peak at the substrate–layer interface as well as intrusion into the layer.

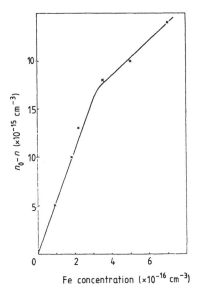

Figure 5.20   Plot of carrier removal against iron concentration for diffused specimen (from Brozel *et al* 1983).

5.4   Cobalt in GaAs

Diffusion of this element was studied by Khludkov *et al* (1972) at $900\,^{\circ}$C and $1000\,^{\circ}$C. The experiments were carried out for a range of ambient arsenic pressures and for diffusion times of 12 hours and 24 hours. After diffusion the samples were chemically etched in cross-section and p–n junctions were revealed. It was found that the junction depth increased with increasing arsenic vapour pressure, although the results found for $900\,^{\circ}$C were not very reproducible. The diffusion coefficient at $1000\,^{\circ}$C and 1 atm. pressure was calculated using the technique described in section 2.4 and a value of $(1.5\text{--}2.3) \times 10^{-11}$ cm^2 s^{-1} was found. As noted in Chapter 2, the method assumes that the profile is an erfc curve and also requires a guess to be made about the surface concentration (they assumed a value of 1×10^{18} cm^{-3}). The quoted diffusion coefficient must therefore be taken as an approximate value.

5.5 Iron and chromium in InP

For InP, iron is the most important of the transition metal dopants, since semi-insulating material is commonly prepared by doping the semiconductor with the element during crystal growth. Radiotracer profiles have been produced by Shishiyanu *et al* (1977) over the temperature range 720–940 °C and by Brozel *et al* (1981, 1982) at 900 °C. In both cases, n-type substrates were employed. The first of these groups plated radioactive iron onto the surface of a slice prior to a diffusion anneal. The profiles indicated deep penetration into the slice, showing a shape rather similar to the 'double profile' characteristic of zinc in GaAs (see, for instance, figure 4.1). They fitted erfc profiles to the bulk sections of the curves and produced values of 'diffusion coefficient' in the region of 10^{-9}–10^{-10} cm^2 s^{-1} for the temperature range employed.

Brozel *et al* (1982) produced profiles which resembled the Cr/GaAs curves of figure 5.1 after a diffusion time of one hour. In their experiments, diffusion took place from the vapour phase and this difference in experimental conditions probably explains the discrepancy between their results and those of the Russian workers. There was no disagreement in interpretation of results, however; in both cases a substitutional–interstitial mechanism was proposed. Both groups of workers used C–V measurements to monitor the electrical effect of the diffused iron, and quite different results were reported. Shishiyanu *et al* found a p–n junction just below the surface, while Brozel *et al* were unable to detect junctions in any of their experiments and, indeed, reported that the presence of iron apparently had no effect at all on the electron concentrations of the samples. They suggested that the earlier results could be attributed to contaminants, possibly introduced at the plating stage.

Brozel *et al* (1982) carried out similar experiments on the Cr/InP system, with very similar results. The chromium penetrated an entire specimen after a one hour anneal at 850 °C, but made no difference to the original electron carrier concentration of the n-type substrate. Figure 5.21 shows the atomic profile, the electron profile and also the starting doping level. At first sight, it might seem worth proposing that for some reason no chromium entered the sample during the diffusion anneal. This approach cannot be sustained, however, since radiotracer measurements are quite unambiguous and there is no doubt that the radioactive isotope of chromium was detected inside the slices. The same can be said of the Fe/InP result described above.

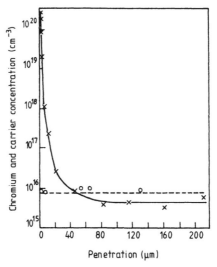

Figure 5.21 Diffusion of Cr into n-type InP for 1 h at 850 °C. Curve shows diffusion profile, circles give electron carrier concentrations taken after diffusion. Broken line gives original carrier concentration (from Brozel *et al* 1982).

It is generally agreed that iron-doped InP, when grown from the melt, becomes semi-insulating because the species Fe_{In} introduces a deep acceptor. The conclusion to be drawn from the work described above is that while iron certainly diffuses into InP, much of it does not sit on the indium site, at least in any simple fashion. A similar conclusion can be drawn for chromium-diffused material. There is evidence elsewhere in the literature that the incorporation of iron into InP is not always straightforward. Messham *et al* (1982) grew layers of InP by liquid phase epitaxy using iron-doped melts. Hall effect and C–V measurements indicated that all layers were n-type, with electron concentrations in excess of 10^{16} cm^{-3}; no semi-insulating or p-type behaviour was observed. The donor and acceptor densities were similar to those obtained from nominally undoped solutions, and the conclusion was drawn that electrically active iron is not incorporated to any significant extent in such layers. They also grew layers from chromium-doped melts and obtained much the same result. Debney and Jay (1980) carried out mobility measurements on semi-insulating InP (Hall effect cannot be used for semi-insulating material, so a magneto-resistance technique was employed) and estimated the concentration of electrically active iron atoms. In addition they performed chemical analyses of the specimens and, as a result, concluded that the concentration of iron atoms exceeded that of electrically active atoms by at least an order of magnitude.

The out-diffusion of iron from as-grown InP substrates was studied at 650 °C by Holmes *et al* (1981). Heat-treated samples were investigated using photoluminescence and SIMS. Their profiles showed an accumulation layer about 0.2 m thick at the surface, similar to the layers reported by many workers for the Cr/GaAs system. Similar results were reported by Oberstar *et al* (1981). Holmes *et al* also grew InP layers on iron-doped substrates by liquid phase epitaxy. SIMS was used to determine iron profiles through the grown layer and into the substrate, as shown in figure 5.22. They found that these profiles fitted reasonably well to the equation

$$C(x,t) = \tfrac{1}{2} C_s \text{ erfc } \frac{x}{2(Dt)^{1/2}} \qquad (5.36)$$

where $C(x,t)$ is the concentration of iron, C_s is the concentration in the bulk of the substrate and x is the distance from the substrate–layer interface. They carried out this experiment for five layers grown at 600–710 °C and were able to plot diffusion coefficients determined from equation (5.36) as a function of temperature. They were found to follow the relationship

$$D = 6.8 \times 10^5 \exp \left(\frac{-3.4(\text{eV})}{kT} \right) \text{ cm}^2 \text{ s}^{-1}. \qquad (5.37)$$

Some care must be exercised in utilising the results summarised in equations (5.36) and (5.37), however. Equation (5.36) is a solution to Fick's law for diffusion between two semi-infinite bars, one of which is homogeneously doped at the start of the experiment (see equation 2.23). This does not seem to be a very good model for describing growth of an epitaxial layer, in which the layer is always thin and, in addition, there is a moving boundary. Even with an improved description of the physical situation, it is extremely unlikely that the profile would be of the form of one of the straight-forward solutions to Fick's law since, as noted above, the diffusion mechanism is almost certainly of the substitutional-interstitial variety. The fact that the experimental points fit equation (5.36) reasonably well is not too surprising. The concentration levels measured cover only one order of magnitude in figure 5.22; many mathematical expressions could be fitted to such a narrow range of results. The value of equations (5.36) and (5.37), therefore, is that they give some idea of the behaviour of iron in InP under these particular conditions, which are fairly typical for LPE growth.

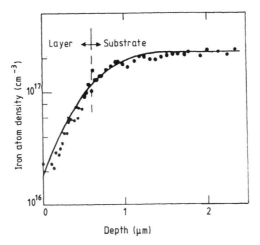

Figure 5.22 Out-diffusion of iron into epitaxial layer. Experimental points are fitted to equation (5.36), and the full curve shows the theoretical result (from Holmes *et al* 1981). Reproduced by permission of *Am. Inst. Phys.*

Similar work was carried out by Chevrier *et al* (1980) who used VPE to grow InP onto substrates which contained iron to a level of 1×10^{16} cm^{-3}. They also used SIMS to demonstrate out-diffusion into the growing layer. In this case the iron level in the epitaxial layer was almost as great as the original substrate level, and a peak occurred at the layer–substrate interface. The profile did not correspond to any erfc expression such as equation (5.36).

Chapter 6

Other Fast Diffusers

Interest in the group I diffusants gold, silver and copper has been somewhat spasmodic over the years. In the early days of GaAs, diffusion of copper was blamed for virtually every experiment that went wrong or device that did not work. This role seems to have been usurped by the transition elements in recent years. Silver has been of interest as a contacting metal for transferred-electron devices (Colliver *et al* 1974, Gray *et al* 1975, Rideout 1975). The fact that they all diffuse very quickly and have a tendency to migrate preferentially to dislocations has led to their use in 'dislocation decoration' studies. In this type of work, group I precipitates actually render dislocations lines visible under infra-red illumination. There is relatively little information on diffusion properties, although the very high rates must clearly be associated with mechanisms which are interstitial in nature. A good deal of the published work is rather old.

6.1 Silver in InP

6.1.1 Diffusion from an external phase

Radioactive silver was diffused into InP by Arseni and Boltaks (1969) over the temperature range 500–900 °C for diffusion times of five minutes to two hours. The diffusion source was a thin layer of the metal deposited either chemically or electrolytically on the semiconductor surface. The general form of the diffusion profiles was very similar to that of the transition elements described in Chapter 5. A high concentration at the surface, C_0 say, drops to an almost homogeneous bulk distribution, C_B, which extends hundreds of microns into the sample. Arseni and Boltaks fitted erfc curves to the interior parts of the distributions and reported an expression for the 'bulk' diffusion:

$$D = 3.60 \times 10^{-4} \exp \left(\frac{-0.59 \pm 0.03 \text{ (eV)}}{kT} \right) \text{ cm}^2 \text{ s}^{-1}. \quad (6.1)$$

On the basis of the evidence presented in the paper, it is not easy to see how this apparently precise result was obtained. In none of the profiles does the bulk concentration drop by more than a factor of two and in most of them, the distribution appears to be virtually horizontal.

The range of experimentation was increased by Tuck and Jay (1978), who performed experiments down to $250\,^\circ$C and still found penetration extending hundreds of microns below the specimen surface after a one hour diffusion. The largest body of experimentation, however, is that produced by Chaoui and Tuck (1983a) who carried out radiotracer diffusions in the temperature range 550–$900\,^\circ$C for diffusion times between 2 minutes and 97 hours. The experiments were performed using metallic silver in the ampoule as a vapour source. Most of the following section will be concerned with the several papers on Ag/InP produced by these workers. Figure 6.1 is taken from their work and demonstrates the evolution of the profiles with time. Note that two of the four curves in the figure show a 'dip' feature just below the surface; this feature, which was also reported by Tuck and Jay (1978), was found to appear unambiguously in certain profiles, although not in a reproducible fashion. The point is demonstrated clearly in figure 6.2, which shows two nominally identical experiments, one giving the dip effect, one not. The values of C_0 and C_B agree well for the two profiles, however.

Variation of the weight of silver used in the ampoules made no discernible difference to the profile obtained. Determination of C_0 was difficult because of the sharp drop in concentration close to the surface, and it was therefore necessary to take a large number of measurements in this region. It was reported that both C_0 and C_B were functions of diffusion time. Surface concentration increased with diffusion time for all temperatures, saturating after a few hours. The behaviour of the bulk concentration was a little more complex. For temperatures less than $800\,^\circ$C, C_B varied in a similar

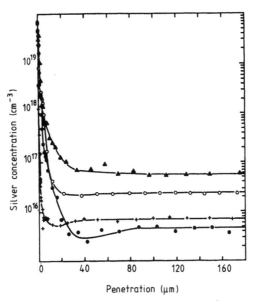

Figure 6.1 Profiles of silver in InP at $850\,\dot{C}$, showing evolution with time. Curves: •, 2 min; +, 15 min; ○, 1 h; ▲, 24 h (from Chaoui and Tuck 1983a).

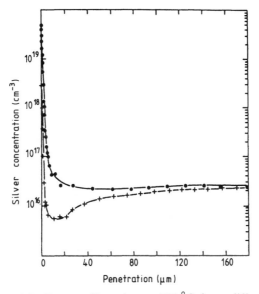

Figure 6.2 Two profiles taken at $700\,^{\circ}C$ for a diffusion time of 24 h. One shows the 'dip' effect (•), the other does not (+) (from Chaoui and Tuck 1983a).

manner to C_0, i.e. it increased with time, but saturated after a few hours of diffusion. For 800 °C and above, however, it continued to increase over the range of diffusion times used.

There is some disagreement concerning the effects of excess phosphorus pressure in the diffusion ampoule. Arseni and Boltaks (1969) reported that excess phosphorus increases the solubility of silver in InP, while both Tuck and Jay (1979) and Chaoui and Tuck (1983a) suggested that both C_0 and C_B are reduced. This difference of opinion may well be a consequence of the difference in the experimental techniques used by the two groups of workers. In the experiments of Arseni and Boltaks, the surface of the InP was effectively protected from the ambient by the layer of plated silver. The interaction of the semiconductor with the phosphorus vapour is likely to be a little more complicated in this case than in an experiment in which the vapour has direct contact with the surface.

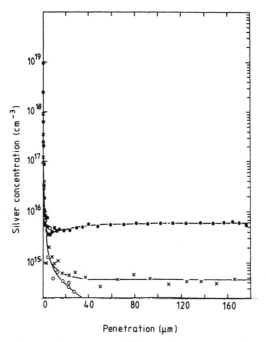

Figure 6.3 Profiles at 550 °C, 24 h, with different amounts of phosphorus in the ampoule. Curves: •, no P; x, 171 μg P; ○, 600 μg P. (from Chaoui and Tuck 1983a).

Chaoui and Tuck found their effect most marked at the lowest diffusion temperatures; figure 6.3 shows three 24-hour profiles taken at 550 °C for different amounts of phosphorus in the ampoule. No value of C_B could be determined for the experiment with the most phosphorus because it is below the sensitivity of the technique employed. For tests carried out at 900 °C, on the other hand, the addition of excess phosphorus affected C_0 and C_B only for the shorter diffusions; after 24 hours, both these quantities were unaffected by the amount of phosphorus added to the ampoule over the range 0–350 $\mu g\,ml^{-1}$. The reductions in C_0 and C_B brought about by the addition of phosphorus were, in fact, greater for short diffusions at all temperatures. This presumably indicates that it takes some time for equilibrium to come about inside the ampoule for this particular system. Close examination of the results showed that although the effect of excess phosphorus on the two concentrations was qualitatively the same, it was significantly greater for C_0.

In all of the above experiments, the InP was nominally undoped; in practical terms, this means that it was n-type at room temp-

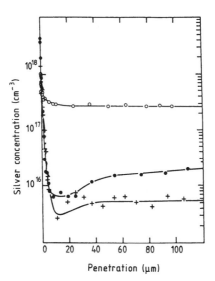

Figure 6.4 Profiles taken at 750 °C, 15 min, using undoped, n-doped and p-doped InP (see table 6.1). Curves: •, undoped; +, n-doped; ○, p-doped (from Chaoui and Tuck 1983a).

erature but intrinsic at most of the diffusion temperatures used. In addition to these experiments, Chaoui and Tuck performed diffusions on InP samples that were doped to a level that would prevent their becoming intrinsic at diffusion temperature. These diffusions were carried out at 750 °C for 15 minutes with no excess phosphorus in the ampoule. The resulting profiles are shown in figure 6.4 together with the corresponding 'undoped' profile. It can be seen that the effect of doping is considerable, with C_B much greater in the p-type material than in the n-type. Full details of these experiments are given in table 6.1.

Table 6.1   Details of diffusion on InP samples experiments doped so as to prevent them becoming intrinsic at diffusion temperature.

Substrate	Dopant	Majority carrier conc. at RT (cm^{-3})	Surface conc. C_0 (cm^{-3})	Bulk conc. C_B (cm^{-3})
n	None	2–5×10^{15}	1.0×10^{19}	2.0×10^{16}
n	Tin	4.5×10^{17}	2.0×10^{18}	5.5×10^{15}
p	Zinc	2×10^{18}	1.3×10^{18}	2.7×10^{17}

6.1.2   A simple theory

The very deep penetration of the silver atoms, taken together with the close similarity with the transition metal profiles shown in Chapter 5, strongly suggest a diffusion mechanism of the substitutional–interstitial type. A relationship between the relevant species of the general form

$$i + v \rightleftharpoons s \qquad (6.2)$$

would then apply, where i is an interstitial atom, v a lattice vacancy and s a substitutional atom. This basic idea was employed by Tuck and Chaoui (1983a) to provide the following qualitative description for their results.

According to this theory, interstitial diffusion is very fast, establishing a homogeneous distribution throughout the crystal. This is reflected in the horizontal portion of the profile at concentration C_B. The surface concentration C_0 represents the concentration of substitutional atoms which is in equilibrium with the external phases within the ampoule and this higher concentration diffuses into the crystal at a rate determined essentially by the vacancy diffusion coefficient, since vacancies must be transported into the bulk of the crystal to fuel equation (6.2). The fact that it takes some time to reach the final value of C_0 is an unusual feature of the Ag/InP system and it can reasonably be assumed that the equilibrium between phases is also taking that time to come about. No ternary phase diagram for In–P–Ag has yet been determined, but from the binary diagrams for Ag–In (Campbell *et al* 1970) and Ag–P (Vogel *et al* 1959), it is clear that a variety of compounds can form between the three elements. Any analysis based on the idea that the silver and phosphorus vapour pressures inside the ampoule are simply those that would be obtained from the individual elements must therefore be discounted.

For temperatures below 800 °C the same explanation for the time dependence of C_B can be invoked. The surface interstitial concentration changes until equilibrium is achieved at the surface, at which time it saturates. Because of the very high diffusion rate of interstitials, this equilibrium concentration is communicated throughout the crystal. It will be suggested below that at these temperatures the silver is virtually all interstitial within the bulk i.e. equation (6.2) does not operate to any great extent because of the lack of a source of vacancies. At 800 °C and above, however, there is no sign of the bulk concentration saturating; it is still increasing after 24 hours. It seems likely that for these higher temperatures there is sufficient thermal energy available for vacancies to be created so that interstitial atoms can join the lattice continuously at a rate determined by the vacancy generated rate.

Figure 6.4 demonstrates that doping present in the original sample can have a strong influence on the resulting profile. Taking a value of 8×10^{16} cm^{-3} for the intrinsic carrier concentration at 750 °C (Casey 1973), the ratios of holes in the three samples, in the order n-type : undoped : p-type are 0.2 : 1 : 25. Table 6.1 shows the ratios for the bulk concentrations in the samples to be in approximate agreement at 0.3 : 1 : 14. This result can be explained by a simple model if three assumptions are made:

(a)   The bulk concentration C_B is made up almost entirely of interstitial silver atoms at 750°C.

(b)   The atoms diffuse as the ion Ag_i^+. This seems reasonable, by analogy with the known diffusion of the acceptor zinc in both GaAs and InP (see Chapter 4).

(c) C_B represents a homogeneous concentration of interstitial atoms extending throughout the crystal, including the surface.

At the surface the equilibrium between the interstitial atoms and the gaseous silver can be written

$$Ag_{vap} \rightleftharpoons Ag_i^+ + e$$

where Ag_{vap} represents a silver atom in the vapour phase and e is an electron. The law of mass action can be applied at equilibrium to obtain

$$P_{Ag} = K[Ag_i^+]n \qquad (6.3)$$

where K is a constant, n is the electron concentration and P_{Ag} is the external vapour pressure of silver. Using the fact that

$$np = n_i^2 \qquad (6.4)$$

we have

$$[Ag_i^+] = \frac{P_{Ag}p}{Kn_i^2} \qquad (6.5)$$

i.e. $[Ag_i^+]$ is proportional to the hole concentration, p. If $[Ag_i^+]$ is identified with C_B, the experimental result is obtained.

The effect of introducing extra phosphorus into the diffusion ampoule must be to bring about changes in the various condensed phases inside the ampoule. This in turn will affect the vapour pressures of phosphorus and silver. It seems reasonable to assume that the phosphorus pressure will increase; the effect on the silver pressure, however, is not obvious. The main effect of the extra phosphorus pressure is to change the stoichiometry of the InP surface (compared to the stoichiometry during an experiment with no extra phosphorus). This change will diffuse into the crystal during the heat treatment, but rather slowly, since its rate will be determined by one of the self-diffusion coefficients in InP, both of which are expected to be small. It is fairly easy to show that an increase in phosphorus pressure brings about an increase in the indium vacancy concentration and a corresponding decrease in the phosphorus vacancies at the surface (see section 2.6).

Consider first the effect of the excess phosphorus on the value of C_B. At the lower diffusion temperatures a reduction in this quantity was observed. This cannot be due to the reduction in phosphorus vacancies at the surface, since the new stoichiometry will not have been transmitted to the centre of the specimen. According to the model proposed above, the bulk concentration would not, in any case, be expected to be sensitive to vacancy concentration since at the lower temperatures the bulk concentration is virtually all interstitial and, according to equation (6.5), this depends only on the external vapour of silver. One is led to the conclusion, therefore, that the addition of extra phosphorus changes the phases inside the ampoule in such a way as to reduce the silver pressure.

It remains to explain why this strong dependence of C_B on added phosphorus is lost at higher diffusion temperatures. It was suggested that at these temperatures equation (6.2) can operate so that silver in the bulk of the specimen can become substitutional. Under the action of this process, most of the existing vacancies become filled with silver atoms early on in the diffusion. If the crystal is able to generate further vacancies, C_B can increase with time at a rate approximately equal to the vacancy generation rate. This corresponds quite well to the situation reported for temperatures in excess of $800\,^{\circ}$C.

For all experiments except those at $900\,^{\circ}$C for 24 hours, the surface concentration C_0 was reduced by the addition of phosphorus to the ampoule. In all cases, the reduction in C_0 was much larger than the reduction in C_B, so the fall in silver pressure which has been postulated as the cause of the behaviour of C_0 cannot alone explain the fall in C_0. It would appear that the change in stoichiometry brought about by the excess phosphorus pressure also acts in such a way as to reduce C_0. If the substitutional silver at the surface occupies a simple lattice position, this argument would seem to suggest that it exists on a phosphorus site. This would be a surprising conclusion; a group I impurity in a III–V semiconductor would be expected to occupy the group III site rather than the group V. Essentially the same situation exists as that described in section 4.4 for the Zn/InP system, and Chaoui and Tuck proposed the same solution, namely that the silver occurs on the lattice as the complex $(V_pAg_{In}V_P)$. As shown in Chapter 4, this gives the desired phosphorus vapour dependence.

A proposal was also made to account for the dip effect shown in figure 6.2. It was suggested that it is associated with the cooling of the ampoule at the end of the diffusion. At this stage, the ampoule wall becomes cool, condensing the silver and phosphorus vapours, but probably leaving the InP specimen close to diffusion temperature for a short time. This leads to a rapid out-diffusion of interstitial atoms from the crystal, and hence depletion just below the surface. Such a process would not be very reproducible, since it depends on the precise details of the quenching process.

6.1.3 Phases present during diffusion

In an extension to the above work, Tuck and Chaoui (1983a) made a study of the changes in the InP slice and in the silver source which took place during the diffusion. At the beginning of an experiment, both were clean and shiny. The slice had flat surfaces, devoid of any features. The differences observed at the finish of an experiment depended on the temperature of diffusion and also on its duration. At lower temperatures (less than 800 °C), for example, whether extra phosphorus was added to the ampoule or not, the InP retained its original appearance, as did the silver. The same result was found at 800 °C provided the diffusion time did not exceed 15 minutes. For longer diffusions, when no phosphorus was added, the InP surfaces became dull, although no change was observed in the silver source. A photomicrograph of a typical InP surface for a diffusion carried out in the absence of excess phosphorus is shown in figure 6.5. The surface is covered with rectangular pits that are lined up in crystallographic directions on the (100) surface used in this work. An enlargement of a single pit is shown in figure 6.6. This result was changed if phosphorus was added to the ampoule; in every case a very clean shiny InP sample with no surface features was seen at the end of the diffusion but the silver source, while retaining its original shape, became black.

At the higher temperatures of 850 and 900 °C the diffusion anneal could have dramatic effects on both the sample surface and the silver source. The surfaces of the semiconductor became matt and pitted unless extra phosphorus was added. In the presence of phosphorus, the diffusion had no noticeable effect on the appearance of the InP. A change occurred in the silver source whether or not extra phosphorus was added, however. From being a flat lump before diffusion, the source became a hard black ball at the end of it, and it was assumed that the source had become molten at diffusion temperature.

Those slices which were diffused in the absence of excess

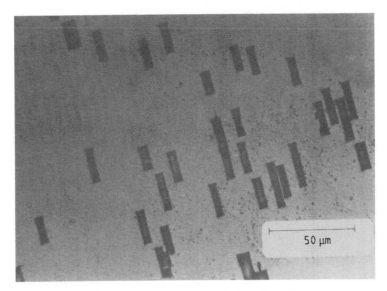

Figure 6.5 Surface damage of InP after silver diffusion at 800 °C with no excess phosphorus in ampoule (from Chaoui 1981).

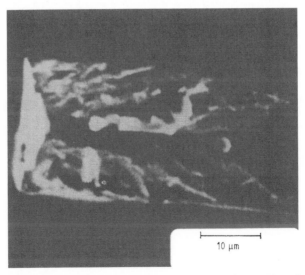

Figure 6.6 Enlargement of a single pit (from Chaoui 1981).

phosphorus were usually found to have lost a small amount of weight during the diffusion, amounting to a micron or less from each surface. The sources were always found to have increased in mass. The mass increased with time of diffusion and the increase was greater for the heavier sources. It is interesting, however, that for a given diffusion time and temperature the mass change ΔM increased sublinearly with the original mass M. This would be expected if the increase were due to a reaction taking place at the surface of a source, since the mass of the source increases as the cube of the linear dimensions, while the surface area increases only as the square. The results of X-ray microprobe analysis of a number of sources are shown in table 6.2. This technique gives a semi-quantitative analysis of the composition of a sample within about 1 μm of the surface. It can be seen that in every case indium and phosphorus were detected in addition to silver. For nearly all the specimens the phosphorus signal was very much greater than that due to indium, however. If the assumption is made that virtually all the increase in mass of the source was due to the addition of phosphorus atoms in some form, it is possible to calculate the ratio of phosphorus atoms to silver in each of the sources at the end of the experiments. On this basis, it was decided that the likely final form of a source was AgP_2 at 750 °C and AgP_3 at 800 °C; both of these compounds appear in the Ag–P binary phase diagram.

Table 6.2   Results of the microanalysis of silver source after diffusion.

Diffusion temp. (°C)	Diffusion time (h)	Elements detected[*]		
		In	P	Ag
700	1	/	//	//
750	1	//	//	//
800	1	T	//	//
850	1	T	/	//
900	1	T	/	//

[*] Intensities of the X-ray lines are denoted as follows: //, strong; /, weak; T, trace.

The source was liquid at diffusion temperatures of 850 and 900 °C despite the fact that the melting point of silver is 961 °C. The Ag–P phase diagram shows a liquid phase appearing in excess of about 880 °C (Vogel *et al* 1959). It is likely that this is essentially the phase observed in this work, presumably with the addition of a small amount of indium which has the effect of lowering the melting point a little.

The features on the surface of the diffused InP were similar to those which had been reported previously for InP heat-treated in a hydrogen ambient (Davies *et al* 1978) and also after diffusion with zinc for certain experimental conditions (Zahari 1976). They were identified as thermal etch pits, filled with indium in the first case and with an indium/zinc phase in the second. It seems likely that a similar explanation is appropriate for the Ag/InP system. When the semiconductor is heated, a very high phosphorus overpressure is required for equilibrium and this can only come from the surface of the sample. The element evaporates, leaving behind liquid indium on the surface. The addition of excess phosphorus eliminates the necessity for the semiconductor to provide the overpressure, no evaporation takes place and the original high quality surface is maintained.

6.1.4 *Electrical measurements on diffused samples*

Electrical measurements were carried out by Tuck and Chaoui (1983b) on diffused InP samples in order to establish the electrical effects of diffused silver on the semiconductor. Before carrying out the tests on the diffused specimens, it was first necessary to establish whether the associated heat treatment caused any changes in the electrical properties of InP. A number of slices were therefore heat treated in the range of 700–900 °C for 15 minutes to 24 hours. In all cases the samples retained their original n-type character after annealing, except for a region of high resistivity which developed in the few microns below the surface. In the bulk of a sample, no change was observed in the carrier concentration within the accuracy of the measurements. Similar experiments have been performed by other workers with much the same result (Lum and Clawson 1979, Rumsby *et al* 1980).

Two techniques were used to measure the electrical properties of the diffused slices. In the first, capacitance–voltage tests were carried out on the Schottky barrier made between a specimen surface and a column of mercury (see section 2.4). Once again a region of high resistivity was found to develop just below the original surface of the diffused slice and it was not possible to make a good Schottky contact until several microns had been removed. Layers were then etched from the slice and C–V measurements were made on each new surface so that a carrier profile could be plotted and compared with the corresponding silver atom profile. The two types of profile are shown in figure 6.7 for a diffusion at 700 °C; also shown is the original electron concentration before diffusion. It can be seen that the specimen remained n-type, with the electron concentration at the pre-diffusion level throughout the bulk of the specimen. The region of high resistivity close to the surface of the samples could not be related to silver, since it appeared in both diffused and heat-treated slices. It seems likely that it was due to some slower-diffusing contaminant, possibly originating in the silica ampoule.

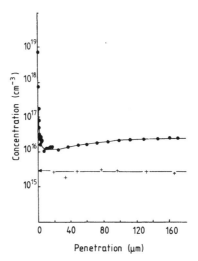

Figure 6.7 Silver profile and electron profile for InP specimen diffused at 700 °C (from Tuck and Chaoui 1983b).

Because of the rather unusual profile shape, it was possible to prepare InP that was homogeneously doped simply by etching about 40 μm from the surfaces of a diffused specimen. Slices were prepared in this way and standard Hall effect measurements were carried out. Table 6.3 shows the results for specimens diffused at

700 °C and 800 °C as well as the original n-type concentration before diffusion. Once again the presence of silver made no significant difference to the electron concentration, although a reduction in mobility was observed.

Table 6.3 Effect of silver diffusion on electrical properties of InP.

Sample treatment	Type	Carrier density (cm^{-3})	Mobility $(cm^2\,V^{-1}\,s^{-1})$	Homogeneous Ag level (cm^{-3})
Untreated	n	5.0×10^{15}	4800	—
Ag-diffused at 700 °C	n	6.8×10^{15}	2430	2.0×10^{16}
Ag-diffused at 800 °C	n	4.9×10^{15}	1910	9.0×10^{15}

The simplest prediction one might make for the electrical effect of a group I impurity in a III–V semiconductor is that it should sit on the group III site and act as a deep acceptor. The above results show quite clearly that this does not occur in Ag/InP, at least in the bulk region of a diffused sample. In each of the experiments the final silver concentration was greater than the original shallow donor concentration. If most of the silver atoms had produced deep acceptors, therefore, the InP would have become p-type and possibly high-resistance, depending on the depth of the acceptor. The result is consistent, however, with the proposal outlined in section 6.1.2 that the bulk concentration C_B is made up almost entirely of interstitial silver atoms. These atoms would be expected to act as deep donors rather than as acceptors. It is fairly easy to show, using Fermi–Dirac statistics, that the addition of deep donor levels to the InP would make very little difference to the free electron concentration i.e. the Fermi level would continue to be determined by the shallow levels. This remains true even when the concentration of the deep levels exceeds that of the shallow levels by more than an order of magnitude.

6.1.5 Out-diffusion of silver

Out-diffusion experiments were performed by Arseni and Boltaks (1969) and Tuck and Chaoui (1984). In both cases, homogeneous specimens were prepared from diffused slices by etching off the high concentration regions close to the surface, as described in section 6.1.4. These new specimens, doped to a level C_B, were then subjected to heat treatments. Two slices were diffused so that profiles could be presented showing the original distribution as well as that after the out-diffusion anneal. Both sets of workers reported out-diffusion profiles which were of the same form as those found for the in-diffusions, but with reduced values of C_B. In general, the C_0 values were not very different to the original; figure 6.8 shows both the original diffusion profile and the out-diffusion profile for 700 °C. It can be seen from the figure that the homogeneous sample was prepared by etching 40 μm from the surfaces of the original diffused slice.

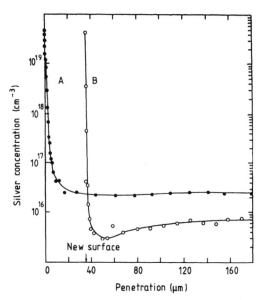

Figure 6.8 Curve A: control diffusion, 700 °C for 24 h. Curve B: as curve A, with about 40 μm etched from the surface followed by an anneal at 700 °C for 15 min (from Tuck and Chaoui 1984).

Tuck and Chaoui (1984) explained their results in terms of the model outlined in section 6.1.2. According to this model, the silver atom exists either as an interstitial which is charged at diffusion temperature or as a lattice atom, in which case it forms the complex

($V_P Ag_{In} V_P$). The detailed form of equation (6.2) therefore becomes

$$Ag_i^+ + e + V_{In} + 2V_P \rightleftharpoons (V_P Ag_{In} V_P). \qquad (6.6)$$

In the bulk this reaction can move to the right only very slowly because of the large numbers of vacancies of both types which are required to participate, so inside the semiconductor all of the silver is interstitial, as noted earlier. Very close to the surface, however, there is a large supply of vacancies and equation (6.6) can operate to form the complex.

When the region of high concentration is removed prior to the out-diffusion part of the experiment, a new surface is revealed. The silver concentration at the surface is C_B. As soon as the heat-treatment commences the concentration of interstitials at the surface can change due to the following phenomena:

(a) loss to the ampoule
(b) loss due to the operation of equation (6.6) i.e. due to creation of the complex
(c) gain by diffusion from the bulk
(d) gain from dissociation of the complex
(e) gain from the external silver vapour in the ampoule (created by out-diffusing silver).

Initially only the first three will be important. Equation (6.6) will be driven strongly to the right because of the large concentration of interstitial atoms and the ready supply of vacancies at the surface. The total concentration of silver at the surface will therefore increase, although the interstitial concentration would be expected to decrease. The interstitial concentration in the bulk will also decrease due to the fast diffusion of interstitial atoms to the surface to facilitate the operation of process (c). The measured profile of silver is therefore the sum of two curves: a shallow substitutional profile and an interstitial profile which have, respectively, a maximum and minimum at the surface. The first of these dominates close to the surface and the second dominates in the bulk.

6.2 Silver in GaAs

At first sight, the results which have been produced for the Ag/GaAs system appear to follow the same pattern as those for

Ag/InP outlined above, and it might be expected that much the same story would suffice for both. In fact, this proves not to be the case; indeed it is surprising how often results from GaAs and InP turn out differently (compare, for example, the diffusion of zinc in the two semiconductors in Chapter 4).

6.2.1 Radiotracer results

The first report on diffusion of silver in GaAs was presented by Rybka *et al* (1962). Radiotracer silver was plated onto the polished semiconductor and diffusion anneals were performed over the temperature range 600–1000 °C. The resulting profiles looked very similar to those already shown for Ag/InP; a high surface concentration dropped sharply to a bulk distribution which extended deep into the crystal. Values for diffusion coefficients were calculated by fitting standard expressions to the bulk portions of the curves, ignoring the surface region, and a relationship was obtained:

$$D = 3.90 \times 10^{-11} \exp \left(\frac{-0.33 (\text{eV})}{kT} \right) \text{ cm}^2 \text{ s}^{-1} \qquad (6.7)$$

Rather similar work was carried out by Boltaks and Shishiyanu (1964) who extended the range of temperatures covered to 500–1160 °C. They found that within 500–800 °C it was possible to fit an erfc curve to the surface portion of the diffusion profile. For higher temperatures this fitting could not be done and it was suggested that evaporation of the sample during diffusion had affected the form of the profiles. Error function complements were also fitted separately to the bulk distributions, and this led to a determination of two diffusion coefficients for each profile: a slow one, D_s, corresponding to the surface region, and a fast one, D_f, for the bulk. The temperature dependencies of the two are given as

$$D_s = 2.5 \times 10^{-3} \exp \left(\frac{-1.5 \pm 0.1 \ (\text{eV})}{kT} \right) \text{ cm}^2 \text{ s}^{-1} \qquad (6.8)$$

$$D_f = 4 \times 10^{-4} \exp \left(\frac{-0.8 \pm 0.05 \ (\text{eV})}{kT} \right) \text{ cm}^2 \text{ s}^{-1}. \qquad (6.9)$$

Note that equations (6.7) and (6.9) refer to the same 'fast' component of the curves. The rather large discrepancy between the two expressions underlines the reservations which have been expressed several times already concerning this way of analysing diffusion results.

Boltaks and Shishiyanu proposed that two different diffusion mechanisms operated, corresponding to the fast and slow components of their profiles. Close to the surface, it was suggested that the high vacancy concentration permitted a vacancy mechanism. This would be expected to be slow, especially in view of the fact that the covalent radius of the silver atom is significantly larger than that of either gallium or arsenic. They suggested that in the bulk of the slice movement of the silver is by interstitial migration of Ag_i^+ ions.

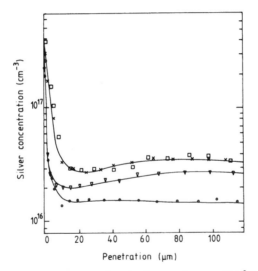

Figure 6.9 Profiles of silver in GaAs taken at $1000\,^\circ$C for different diffusion times, ○, 30 min; ▽, 45 min; x, 60 min; □, 240 min (from Tuck and Adegboyega 1980)

Tuck and Adegboyega (1980) carried out a series of diffusions at $1000\,^\circ$C demonstrating the progression of profile shape with time. Figure 6.9 shows a number of such profiles for diffusion times between 30 minutes and 4 hours. Note that the 'dip' effect appears, as in the Ag/InP profiles. The approximately constant value for the surface concentration, C_0, indicates that surface equilibrium was achieved early in the diffusion process, in contrast to the Ag/InP system. The bulk concentration, C_B, on the other hand, builds up with time to a maximum value of 3.4×10^{16} cm^{-3}. This is shown clearly in the figure; the value of C_B for the 4-hour diffusion is the same as that for the 1-hour experiment.

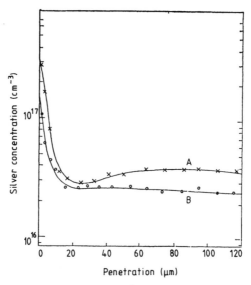

Figure 6.10   Profiles at 1000 °C for different weights of arsenic in the ampoule. Curve A, no As; curve B, 3 mg As (from Tuck and Adegboyega 1980).

The effect of adding arsenic to the ampoule is shown in figure 6.10. There is a reduction in both the surface and bulk atom concentrations, but the penetration characteristics are unchanged. The effect was reported as being rather insensitive to the amount of arsenic added to the ampoule. This result was qualitatively the same as that found for silver in InP and it presented the same problem of interpretation. The obvious explanation is that the silver atom occupies the group V site, but this would appear most unlikely in view of the electrical measurements to be described in the next section. Tuck and Adegboyega opted for the same explanation as that given in section 6.1.2, namely that the addition

of extra arsenic to the diffusion ampoule has the effect of reducing
the silver over-pressure and that this reduction is sufficient to
out-weigh any effect due to increased arsenic vapour pressure.

6.2.2   *Electrical measurements on diffused specimens*

Shishiyanu and Boltaks (1966) carried out Hall effect measurements as a function of temperature in order to determine the energy
levels associated with diffused silver in GaAs. The measurements
were taken over the range 77–770 K. The starting material was
n-type, with electron concentration $1-2 \times 10^{16}$ cm^{-3}. Silver was
diffused into slices of this material at 800 °C to provide samples
saturated to a level of 3×10^{10} cm^{-3} and had the effect of turning
the samples p-type. The conclusion was drawn that introduction of
the silver had given rise to an acceptor level (0.11 ± 0.01) eV above
the top of the valence band. The experiments were somewhat
complicated by the presence of thermal conversion (see section
5.2.2). This was revealed by control experiments in which the GaAs
was annealed in the absence of silver. The slices still turned p-type,
but the Hall measurements showed that this conversion was due
to a different acceptor level, 0.15 eV above the valence band.
Shishiyanu and Boltaks were of the opinion that there might be
other levels due to silver in addition to that at 0.11 eV, rather
deeper in the forbidden gap, but that the thermal conversion
phenomenon prevented their being detected in the Hall experiments.

Tuck and Adegboyega (1980) carried out electrical measurements
on silver-diffused GaAs which were very similar to those described
in section 6.1.4 for Ag/InP. The results, however, were quite
different. All of the samples were p-type after diffusion. Figure
6.11 shows the results of C–V measurements on two specimens,
one diffused with added arsenic, one without. In addition to the
hole concentration profiles, the corresponding atomic profiles are
also shown for comparison. Bearing in mind the thermal conversion problems encountered by the earlier workers, it was important
to confirm that the p-type conductivity demonstrated by the
diffused specimens was due to the presence of silver and not to the
heat treatment associated with the diffusion process. An n-type

substrate was therefore subjected to a heat treatment identical to that involved in a silver diffusion cycle. The slice did indeed convert to p-type, but at a level almost two orders of magnitude below the carrier concentrations observed in the silver-diffused specimens. The carrier profile for this annealed sample is shown as curve A in figure 6.11. It seemed reasonable to assume, therefore, that the p-type conductivity of the diffused specimens was due to the presence of silver; it is worth noting in this context that both hole and atom concentrations show the same dependence on arsenic over-pressure, i.e. the effect of adding arsenic is to reduce the concentrations.

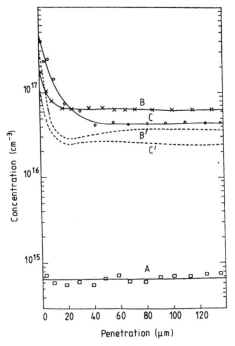

Figure 6.11 Carrier concentration profiles for GaAs diffused at 100 °C for 1 h. Curve B had no arsenic added, curve C had added arsenic. The broken curves B', C' give the corresponding atomic distributions. Curve A gives the concentration of holes due to thermal conversion (from Tuck and Adegboyega 1980).

In further experiments, homogeneous diffused samples were produced by etching about 50 μm from each flat surface of two specimens. Standard conductivity and Hall measurements were carried out to measure hole concentrations and mobilities. The results are given in table 6.4, together with the atomic concentrations and the hole densities derived from the C−V measurements. It can be seen that the C−V results gave carrier concentrations which are rather higher than the atomic concentrations, while the Hall values are rather lower. This led Tuck and Adegboyega to conclude that silver acts as an acceptor with an approximate 1:1 relationship between hole and atom concentrations, in agreement with the Russian workers. The possibility that diffused silver can act as a double acceptor cannot be ruled out, however, on the basis of these results.

Table 6.4

	Silver diffused	Silver diffused + arsenic
Radiotracer, atoms (cm^{-3})	3.6×10^{16}	2.6×10^{16}
Hall effect, carriers (cm^{-3})	2.3×10^{16}	1.4×10^{16}
Mobility, holes (cm^2 V^{-1} s^{-1})	415	559
C−V, carriers (cm^{-3})	6.2×10^{16}	4.5×10^{16}

6.2.3 A diffusion model

In this section a model due to Zahari and Tuck (1983) will be described, in which an attempt was made to reproduce the experimental results of Tuck and Adegboyega (1980). The method is essentially a simplified version of the model presented for the Zn/GaAs system in section 4.2.2. The interaction between interstitials, vacancies and substitutionals is unchanged, so equation (4.21) still applies:

$$KIV = S. \tag{6.10}$$

The assumptions underlying the proposed mechanism can be summarised as follows:

(i) Interstitial impurity atoms diffuse with an effectively infinite diffusion coefficient, giving the condition

$$I = I' \text{ for all } x, \quad t > 0$$

where I' is the concentration of interstitial atoms at the surface. It is this condition which makes the present model rather easier to handle than the previous Zn/GaAs one.

(ii) Substitutional silver atoms have zero diffusion coefficient.

(iii) Vacancies in the bulk are created at a rate proportional to the short-fall, so equation (4.28) still applies.

(iv) Self-diffusion by host atoms takes place by the simple vacancy mechanism described by equation (4.26).

(v) The reaction rate for equation (4.20) is infinitely fast, so equation (6.10) is always satisfied. Bearing in mind assumption (i), above, equation (6.10) can be written in the simpler form

$$\frac{V}{S} = \frac{K}{I'} = K' \tag{6.11}$$

The computing procedures were as outlined in section 4.2.2. A set of calculated profiles is shown in figure 6.12, computed using an iteration time Δt of 1 second and node spacing Δx of 1 μm. The profiles are to be compared with figure 6.9. Reasonable qualitative agreement can be seen between the two sets. The profile shape is very similar, and both show a constant surface concentration C_0 and saturation occurring in the bulk concentration C_B. The approach to saturation is rather faster in the case of the experimental curves, however,

In the calculation, the concentration of sites, L, was taken as the number of gallium sites per unit volume in GaAs: 2.21×10^{22} cm^{-3}. The quantity S' was the surface concentration in figure 6.9 $(4.0 \times 10^{17}$ cm$^{-3})$ and the value used for the self-diffusion coefficient of gallium in GaAs was estimated from the work of Palfrey *et al* (1981). In this model, V' is the equilibrium concentration of gallium vacancies, V_{Ga}; a fitting procedure was used to adjust this concentration, together with vacancy generation constant, k_v, to give a fit between theoretical and experimental profiles. The best-fit values were $V' = 1 \times 10^{16}$ cm^{-3} and $k_v = 2.3 \times 10^{-4}$ s^{-1}.

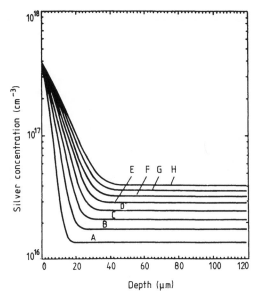

Figure 6.12 Computed profiles for silver diffusion in GaAs. Diffusion times are: A, 30 min; B, 60 min; C, 90 min; D, 120 min; E, 150 min; F, 180 min; G, 210 min; H, 240 min (from Zahari and Tuck 1983).

6.3   Silver in InAs and GaP

Silver was diffused into InAs over the temperature range 450–900 °C by Boltaks *et al* (1967) using both n-type and p-type substrates. Two experimental techniques were used for obtaining diffusion profiles, both involving plating radiotracer silver onto the slices prior to the diffusion anneal. The first method was the normal etching and counting technique for plotting profiles point by point. In the other technique, the sample was sectioned at right-angles to the original surface at the end of the anneal and an auto-radiographic exposure made of the section using X-ray film. The density of blackening of the film was measured by a microphotometer, giving a direct estimate of the diffusion profile. The latter method is much less accurate than the section-and-count technique, but is quicker, and suitable for investigating very thick samples.

Profiles obtained by the first method looked very much like the

profiles described above for silver in InP and GaAs. The dip effect
was quite marked, and the bulk level C_B rapidly reached saturation. To investigate in more detail the distribution in the interior of
the InAs, samples which were several microns thick were diffused,
and the auto-radiograph method was used to determine the profiles.
The bulk sections of the curves fell in concentration by a factor of
up to ten over the depth measured. This permitted the following
expression to be determined for the bulk diffusion coefficient:

$$D = 7.3 \times 10^{-4} \exp \left(\frac{-0.26 \pm 0.05 (\text{eV})}{kT} \right) \text{cm}^2\,\text{s}^{-1}. \quad (6.12)$$

An interesting series of experiments was carried out in which an
external electric field was applied as silver was diffusing in InAs.
The experiments were performed in the range $520-680\,^\circ\text{C}$ with
electric fields of $0.06-0.12\ \text{V cm}^{-1}$. The distribution of silver in the
samples was again determined by the radiographic method. These
tests showed that under the action of the field, silver travelled
preferentially to the cathode, i.e. the metal migrated in the form of
positive ions. Such a result is consistent with the suggestion made
by Boltaks *et al* that the diffusing species is the charged interstitial
Ag_i^+.

Differential Hall measurements were carried out on diffused
specimens, permitting carrier concentration profiles to be plotted
and compared with atomic profiles of silver. The results of these
experiments were something of a surprise. The over-all effect of the
diffusion was to increase the electron concentration of n-type
samples and to convert p-type to n-type, i.e. the silver was acting as
a donor. The profiles for both n- and p-type substrates are shown in
figure 6.13; the original carrier concentrations were $6 \times 10^{16}\ \text{cm}^{-3}$
and $1 \times 10^{17}\ \text{cm}^{-3}$ respectively. Also shown in the figure are the
calculated densities of carriers introduced by the diffusion process.
It can be seen that the distributions of the carrier densities followed
the silver distributions very closely. In the case of n-type samples,
the concentration of carriers introduced by diffusion was less than
the concentration of silver atoms, while in p-type material it was
approximately the same.

The donor activity of silver could be interpreted as indicating
that all of the silver exists in the InAs lattice in the form Ag_i. This
seems unlikely in view of the very high donor concentrations near
the surface in figure 6.13. Boltaks *et al* concluded that silver acts as

a donor even when occupying a lattice position. There seems a strong possibility that this may be due to the formation of a complex of some kind.

A small amount of work has been reported on the diffusion of silver into GaP (Chaoui and Tuck 1983b). This was carried out at 1140 °C for very short diffusion times. Profile shapes were presented which showed very similar characteristics to those which have been presented for silver in the other III–V semiconductors.

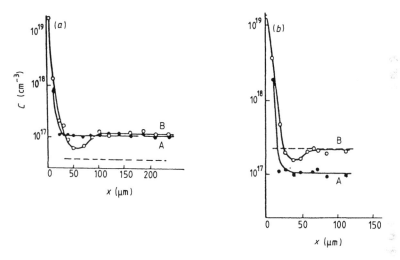

Figure 6.13 (*a*) Silver profile (curve A) and electron profile (curve B) after diffusion into n-type InAs at 886 °C for 1 h. Broken line represents the calculated density of carriers introduced by the diffusion (from Boltaks *et al* 1967). (*b*) Same as (*a*) but for p-type substrate (Boltaks *et al* 1967).

6.4 Diffusion of gold

Information on the diffusion of this element in the III–Vs is rather meagre. Papers, many of them rather short, have presented profiles for diffusion in GaAs, InP, InAs and InSb. In all cases, the profile shapes have been of the same general form as those exhibited by silver and the authors have attributed them to a diffusion mechanism which is primarily interstitial.

Sokolov and Shishiyanu (1964) diffused gold into GaAs over the temperature range 740–1025 °C and found an expression for the bulk diffusion coefficient of

$$D \simeq 10^{-3} \exp\left(\frac{-1.0 \pm 0.2(\text{eV})}{kT}\right) \text{cm}^2\text{s}^{-1}. \qquad (6.13)$$

The pre-exponential term is described as approximate. The authors did not present detailed results in the paper, so it is not clear exactly how approximate this value is. The interesting observation was made from auto-radiograph tests that gold accumulated at twin boundaries in their specimens. No preferential penetration of gold along these boundaries was observed, however. Shishiyanu and Boltaks (1966) continued this work by carrying out electrical measurements on gold-diffused GaAs, using n-type substrates. Once again, the results were somewhat obscured by the presence of thermal conversion (see also section 6.2.2). It was concluded, however, on the basis of Hall measurements, that the gold had introduced an acceptor level 0.09 eV above the top of the valence band.

Diffusion of gold in InAs has been described by Rembeza (1967), covering the range 600–890 °C. He obtained an expression for the bulk diffusion coefficient

$$D = 5.8 \times 10^{-3} \exp\left(\frac{-0.65 \pm 0.05(\text{eV})}{kT}\right) \text{cm}^2\text{s}^{-1}. \qquad (6.14)$$

Experiments were also carried out in the presence of an electric field (see also section 6.3). It was found that the effect was to displace the gold towards the cathode, and it was proposed that the main diffusing species was therefore Au_i^+ by analogy with the work on silver. It is interesting that the 'dip' effect was reported, appearing in a similar manner to the way it occurs for silver in the III–Vs.

The Au/InSb system was studied by Boltaks and Sokolov (1964) within the temperature range 140–510 °C. They found that the bulk section of the profiles followed an expression of the form of equation (6.14) with a pre-exponential term 7×10^{-4} cm^{2}s^{-1} and an activation energy 0.32 eV. A substitutional–interstitial treatment which was essentially a simplified form of the earlier Sturge (1959) theory (see section 5.1.4) was offered by way of explanation. Very similar work on gold in InP was performed by Rembeza (1969) for 600–820 °C who found that the bulk profile followed the same general rule as the other systems, with a pre-exponential value 1.32×10^5 cm^{2}s^{-1} and an activation energy of 0.48 ± 0.01 eV.

6.5   Diffusion of Copper

It is well-known that copper can enter a III—V semiconductor very easily during processing and usually great care is taken to prevent this happening. The point was made very nicely some years ago in a short note from Gansauge and Hoffmeister (1966). They annealed a GaAs wafer inside a closed silica ampoule for two hours at 1050 °C. Using a radioactive method, they established that copper existing as a contaminant in the silica could introduce a doping level of 1×10^{15} cm^{-3} into the semiconductor. They went further. The same ampoule was used for a second and a third anneal, using fresh GaAs wafers each time. Even after the third anneal, the GaAs contained copper to a level of 7×10^{13} cm^{-3}.

  Hall and Racette (1964) carried out radiotracer experiments in which both diffusion and solubility of copper in GaAs were studied. They interpreted their results in terms of a substitutional—interstitial model in which the interstitial occurs as a singly ionised donor and the substitutional atom acts as an acceptor. A number of experiments was performed using doped GaAs. In these tests, the original doping was greater than any concentration of copper introduced during diffusion, ensuring that the Fermi level remained constant throughout the anneal. These tests established that the solubility of copper increases with p-type doping in a roughly linear fashion, a finding which is consistent with the copper being primarily in the interstitial form in p-type material. This can be seen as follows. Consider an experiment in which copper in the vapour, vapour pressure P_{Cu}, is in equilibrium with a copper-doped GaAs slice. The equilibrium at the surface between copper in the vapour and the charged interstitial atom in the crystal can be written

$$Cu_v \rightleftharpoons Cu_i^+ + e \qquad (6.15)$$

which gives rise to the mass-action formulation

$$P_{Cu} = K[Cu_i^+]n \qquad (6.16)$$

or,

$$[Cu_i^+] = \frac{pP_{Cu}}{Kn_i^2}. \qquad (6.17)$$

A similar argument shows that the concentration of substitutional copper should be enhanced in n-type material, and Hall and Racette did, in fact, conclude that the substitutional atom predominates in intrinsic and n-type GaAs. At 700 °C, for instance, the ratio of substitutional to interstitial concentrations in intrinsic material is given as about 30.

The hypothesis of the diffusing interstitial having a positive charge was further tested by carrying out copper diffusions in the presence of an electric field of 2 V cm^{-1} at about 200 °C. Following this treatment the copper profile was measured and it was invariably found to be shifted towards the negative end of the specimen. Detailed calculations based on the shapes of the profiles indicated that the diffusing interstitial had unit positive charge. Hall effect measurements on copper-diffused n-type GaAs indicated that the substitutional copper acts as a double acceptor. This last result has been confirmed by Fuller *et al* (1964) and Allison and Fuller (1965).

An interesting anomaly in the diffusion of copper in n-type GaAs was reported by Larrabee and Osborne (1966), who carried out radiotracer diffusions at 810 °C. Normal diffusion profiles looked very much like those described earlier in the chapter for silver and gold in GaAs, with a relatively high surface concentration and an almost horizontal bulk portion. Several experiments were carried out, however, in which copper was diffused simultaneously with some acceptor such as manganese or zinc. The conditions were chosen to make the surface concentration of this acceptor much greater than that of the copper. An example of the resulting copper diffusion profile is shown in figure 6.14. A large dip is observed just below the surface, resembling the effect previously described for silver and gold, but much larger in magnitude. In these experiments the position of the minimum corresponded closely to the position of the p−n junction created by the accompanying acceptor. It was suggested that the phenomenon was an example of the built-in field effect (see section 2.7), with the positively charged diffusing interstitial copper atom being held up by the p−n junction field. No detailed calculations were presented in support of this proposition, however.

There have also been reports describing the diffusion of copper in InP (Arseni 1969), InAs (Fuller and Wolfstirn 1967), InSb (Stocker 1963) and AlSb (Wieber *et al* 1960). Most of these investigations

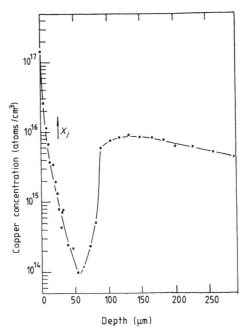

Figure 6.14   Copper profile in GaAs after diffusing simultaneously with manganese at 815 °C for 6 h (from Larrabee and Osborne 1966). Reproduced by permission of *The Electrochemical Society, Inc.*

involved depositing a layer of radioactive copper onto the surface of a semiconductor slice before annealing in a closed ampoule. The resulting diffusion profiles were of the form described above for the Cu/GaAs system, with the exception of those of Fuller and Wolfstirn for InAs, which were fitted to erfc curves. This last result can only be described as surprising; it seems likely that the portion of profile close the surface was not detected by the experimenters for some reason. Most workers fitted erfc expressions to the bulk sections of the curves and hence derived an expression for activation energy. In general, low values were obtained, in the range 0.5–0.7 eV, indicative of diffusion processes which are primarily interstitial. The value 0.36 eV quoted by Wieber *et al* for AlSb looks suspiciously low, however.

Electrical measurements indicated once again that a large proportion of the copper could act as a donor in these materials. In

InAs, the interstitial donor was revealed as the major dopant, turning p-type material n-type. Interestingly, Boltaks *et al* (1968) found that this effect was only temporary. They reported that the samples reverted to the original electrical properties after 8–10 days at. room temperature. This was attributed to copper atoms migrating to dislocations and forming precipitates. Dislocations were also found to have an effect on copper diffusion in InSb. Stocker (1963) reported that for dislocation densities above $10^3 \, \text{cm}^{-2}$ an enhanced diffusion penetration was obtained. The effect was quite large; at $300 \, ^\circ \text{C}$, for example, the effective diffusion coefficient for samples with a dislocation density of $10^5 \, \text{cm}^{-2}$ was almost 10^3 times that for samples with a density of $10^3 \, \text{cm}^{-2}$. This result underlines the important role that dislocations can play as sources and sinks of defects during diffusion.

Chapter 7

Self-diffusion and Related Phenomena

The foregoing chapters have, perhaps, indicated to the reader that impurity diffusion in the III–Vs is often complex and that, in addition, matters can be further complicated by the simultaneous occurrence of self-diffusion. In these circumstances, it might be thought logical to gain a thorough understanding of self-diffusion before allowing a third element into the crystal. Not so. In fact, the amount of reliable self-diffusion data in the literature is minimal; it does not begin to compare, for instance, to the information available on the Zn/GaAs system. The fact is that a quite different logic applies in this subject, as in many others. The work which is carried out is driven by the problems of the day in device manufacture rather than by any desire to build up a body of knowledge in an orderly fashion. Thus the phenomenon of diffusion-induced disorder in the GaAs/AlAs system (dealt with later in this chapter) has been the subject of many papers, while the rather more straightforward problem of self-diffusion in AlAs has hardly been considered. It is difficult to see, however how the first problem is to be solved without having a good understanding of the second. It follows that in much of this chapter we have no option but to make the best of a bad job. In addition to true self-diffusion, the chapter also deals with 'impurity' diffusion of group III and group V elements. As is usually the case, the best available information is on GaAs.

7.1 Diffusion in GaAs

7.1.1 Gallium in GaAs

Radiotracer techniques are very suitable for study of this system, as indeed they are for most self-diffusion investigations, since they make a clear distinction between those atoms which have diffused into the semiconductor and those which were there all the time. The experiment is, nevertheless, one which is rather difficult to carry out in practice. The only suitable gallium isotope, Ga^{72}, has a half-life of only 14 hours, so a high degree of organisation is required in order to obtain results before the isotope dies. Self-diffusion of gallium is also very slow, so it is necessary to carry out long anneals and have a sectioning technique which can characterise a profile which extends only a micron or so into the crystal. The fact that the profiles are so shallow also makes it essential to maintain the integrity of the original surface of the semiconductor slice; it is only too easy to lose several microns from the surface during a long anneal.

The most satisfactory set of data is that produced by Palfrey *et al* (1981). They evaporated radioactive gallium onto GaAs slices and annealed for 24 hours at temperatures in the range 1025–1100 °C. the diffusions were carried out in closed ampoules, and an arsenic over-pressure of 0.75 atm was used to prevent decomposition of the surface. The addition of excess arsenic to the ampoule indicates that the diffusions took place on the arsenic-rich side of the Ga–As phase diagram. Sectioning was carried out by an anodic oxidation method due to Hasegawa and Hartnagel (1975), permitting layers of only 250–500 Å to be removed. A typical profile is shown in figure 7.1. Given the experimental arrangement, it is not immediately obvious how to describe the boundary conditions for the diffusion; Palfrey *et al* found in practice that the profiles fitted Gaussian curves rather well (i.e. the gallium acted as a limited diffusion source) and used these curves to determine a set of diffusion coefficients. The expression

$$D = 3.9 \times 10^{-5} \exp\left(\frac{-2.6 \pm 0.5(\text{eV})}{kT}\right) \text{cm}^2\text{s}^{-1} \qquad (7.1)$$

was found to apply over the temperature range of the experiments.

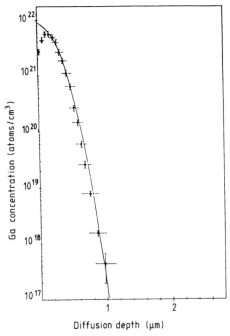

Figure 7.1 Diffusion profile of gallium in GaAs at 1050 °C, fitted to a Gaussian curve (from Palfrey *et al* 1981). Reproduced by permission of *The Electrochemical Society Inc.*

Note that this amounts to very low coefficients, in the range 10^{-14} -10^{-15} cm^2 s^{-1}.

In view of the uncertainty concerning the exact boundary conditions for the experiment, it is of interest to speculate on what phases were likely to be present in the ampoule during a diffusion. The ampoule volume is given as 24 cm^3. In order to obtain the quoted over-pressure of arsenic of 0.75 atm (presumably calculated assuming that all the added arsenic went into the vapour phase), it would have been necessary to have added over 20 mg of arsenic. The mass of radioactive gallium evaporated onto the sample is not given, but it was almost certainly much less than this. The ampoule therefore contained a GaAs slice, an amount of extra arsenic and a very small quantity of extra gallium. In the region of 1100 °C and

on the arsenic-rich side of the phase diagram, GaAs is in equilibrium with a liquid of approximate composition 80 percent arsenic,
20 percent gallium (Panish 1974). This liquid has an equilibrium
vapour pressure of arsenic well in excess of 10 atm (Jordan 1971).
Since there was not sufficient arsenic in the ampoule to produce this
pressure (fortunately so, for it would quite possibly have caused
the ampoule to explode) the liquid would not have formed. At
equilibrium, therefore, the ampoule would probably have contained solid GaAs, gallium vapour and arsenic vapour, but no
liquid phase. The assumption that virtually all of the extra arsenic
went into the vapour phase is therefore seen to be essentially
correct.

Under the circumstances outlined above, the following sequence
of events is likely to occur when an ampoule is inserted into the
diffusion furnace. Initially, all of the arsenic goes into the vapour
phase. Some of the gallium also enters the vapour, but this amount
is very small, since the vapour pressure of the element is only in the
region of 10^{-4}–10^{-5} atm at diffusion temperature. The rest of the
gallium stays on the surface of the slice as a liquid. This liquid then
reacts with the arsenic vapour and a thin GaAs layer is grown on
the surface of the slice. All this occurs on a time-scale which is very
short compared to the over-all diffusion time. It is possible,
therefore, that the diffusion source for the experiments was a thin
epitaxial layer of GaAs, containing radioactive gallium.

The surface concentrations of the profiles produced by Palfrey
et al depended on the amount of gallium evaporated onto the
samples and varied between 3×10^{19} cm^{-3} and 2×10^{22} cm^{-3}. It
should be remembered that a GaAs crystal has a density of gallium
atoms of about 2.2×10^{22} cm^{-3}. It can be assumed, therefore, that
for the profile with a surface concentration of 3×10^{19} cm^{-3}, only
about 0.1 percent of the gallium atoms at the surface were
radioactive at the end of the experiment, the rest being non-
radioactive. Since the surface was in kinetic equilibrium with the
surrounding vapour, both losing atoms to it and gaining atoms
from it, we can assume that, to a first approximation, the gallium
vapour also comprised 0.1 percent of radioactive atoms. All of the
profiles exhibited a fall in concentration at the surface (see figure
7.1). It was suggested that this might be due to a preferential
removal of gallium from the surface. This seems unlikely in view of
the very limited range of stoichiometry permitted to GaAs by the

phase diagram. It is possible that the region showing the fall could correspond to the thin expitaxial layer of GaAs mentioned above. It seems more likely, however, that the drop at the surface is due to the fact that the proportion of radioactive gallium in the external vapour falls during the course of an experiment due to radioactive atoms entering the slice. This leads to a steady reduction in the surface concentration as the diffusion progresses.

Radiotracer profiles were also plotted by Goldstein (1961), at slightly higher temperatures than those employed in the above work. Using electroplated layers of gallium as the diffusion source in the range 1125–1225 °C, he found a relationship of the form of equation (7.1), with a pre-exponential term of 1×10^7 cm^{2}s^{-1} and an activation energy 5.60 ± 0.32 eV. This result is significantly different to that of Palfrey *et al* (1981), who suggested that differences in experimental conditions might explain the discrepancy. It is difficult to see, however, how such a major disagreement can be explained in this manner. Goldstein's very high value of activation energy is particularly surprising. He was of the opinion that gallium diffuses by a substitutional mechanism on the group III sub-lattice; his argument is given in section 7.1.2.

7.1.2 Arsenic in GaAs

In a later paper, Palfrey *et al* (1983) presented a limited amount of data on the diffusion of arsenic in GaAs. Profiles were plotted using radioactive As^{76} vapour as the diffusion source. The technique was essentially the same as that outlined in section 7.1.1, with similar difficulties arising from the short half-life of the isotope of 26.5 hours. Diffusion time was 24 hours and the arsenic vapour pressure used was again 0.75 atm. Three profiles taken at 1000, 1025 and 1050 °C could be fitted by erfcs and gave diffusion coefficients of 5.2×10^{-16}, 8.7×10^{-16} and 1.5×10^{-15} cm^{2}s^{-1} respectively. Palfrey *et al* attributed an activation energy of 3 eV to the diffusion process, but this can only be taken as an approximate value in view of the small quantity of data. It is interesting to note, however, that the values of diffusion coefficient for arsenic were definitely lower than those found by the same workers for gallium. Two further diffusions were performed at an arsenic pressure of 3 atm. In each case, a lower value of diffusion coefficient was found than had been established for the work at 0.75 atm. It was pointed out that the main effect of increasing the arsenic pressure is to suppress the arsenic vacancy concentration (and increase that of gallium vacancies). The result thus suggests that arsenic vacancies are involved in the diffusion mechanism; it was proposed that arsenic diffuses via vacancies on its own sub-lattice.

Arsenic self-diffusion was also studied in the earlier work of Goldstein (1961). Once again, the temperature range used was slightly different to the work of Palfrey *et al* (1983) and once again the results differed significantly. Over the range 1200–1225 °C, Goldstein reported that diffusion coefficient varied with temperature according to the relation

$$D = 4 \times 10^{21} \exp\left(\frac{-10.2 \text{ (eV)}}{kT}\right) \text{ cm}^2 \text{ s}^{-1}. \tag{7.2}$$

For the rest of the temperature range, 1125–1200 °C, the diffusion coefficient was constant at 4×10^{-14} cm^2 s^{-1}. This result is remarkable in two respects. In the first place, the activation energy of 10.2 eV seems unreasonably high. Secondly, the temperature-independent diffusion coefficient below 1200 °C does appear rather implausible. It has been suggested by Kendall (1968) that the value of 4×10^{-14} cm^2 s^{-1} simply represented the precision limit of the sectioning technique employed. Palfrey *et al* proposed, rather generously, that the discrepancy between the two sets of results might be due to a different diffusion mechanism being operative at the slightly higher temperatures used in the earlier work.

Goldstein did, however, agree with Palfrey *et al* in attributing the results to a substitutional diffusion mechanism operating on the arsenic sub-lattice. In support of this opinion, he pointed to the large discrepancy in diffusion constants and activation energies for gallium and arsenic, which would not be consistent with any mechanism of atomic migration via nearest-neighbour vacancies or via ring motion, both of which would involve equal numbers of each species. Each group III atom has four group V atoms as nearest neighbours and *vice versa*. Hence a mechanism involving nearest-neighbour exchange would be expected to give similar activation energies for both species. Unfortunately, this argument cannot be applied to the work of Palfrey *et al*, since this suggested that the two activation energies are indeed similar. The best evidence to date for the sub-lattice mechanism remains the arsenic vapour dependence reported by the later workers.

7.1.3 Phosphorus in GaAs

The ternary compound $GaAs_xP_{1-x}$ exists over the entire range of composition from $x=1$ to $x=0$ and early work (Stone 1962, Goldstein and Dobin 1962) established that it can be produced by diffusing phosphorus into GaAs. Diffusions were carried out by Jain *et al* (1976) over the temperature range 800–1150 °C and for times ranging from 2 to 225 hours. The experiments were performed in an ambient phosphorus pressure of 35 atm. Profiles were determined from reflectivity measurements on the diffused specimens. The shift of the major reflectivity peaks in $GaAs_xP_{1-x}$ as a function of x is well-established (Williams and Jones 1965) and the composition-depth data could be obtained from the positions of the peaks in the reflectance spectra taken at different depths in the diffused region. A typical set of results is shown in figure 7.2.

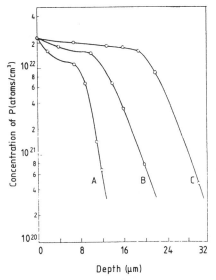

Figure 7.2 Profiles of phosphorus in GaAs: A, 800 °C, 40 h; B, 1000 °C, 40 h; C, 1100 °C, 24 h (from Jain *et al* 1976). Reproduced by permission of *Pergamon Journals Ltd*.

The profiles in the figure are anomalous in the sense that they do not correspond to a standard solution of the diffusion equation. This might be considered surprising in view of the well-behaved nature of the results for arsenic in GaAs, reported above. The following argument may make the result slightly less surprising. The diffusion of radioactive arsenic into GaAs is essentially an 'isoconcentration' experiment. During the diffusion process radio-active arsenic atoms are changing places with non-radioactive arsenic atoms, i.e. to a high degree of approximation, nothing is happening. There is no gradient of arsenic atoms within the crystal and activity of the arsenic is constant (see Chapter 2). The phosphorus case is quite different. One way of looking at $GaAs_xP_{1-x}$ is to consider it as a solid solution of GaP in GaAs. Reference to figure 7.2 shows that the amount of GaP solute varies from 100 percent near the surface to zero deep within the slice. It has been noted in section 2.6.1 that, in general, solutions are not ideal and that equations (2.63) and (2.64) are, at best, only approximations. Work on the III–V solid solutions has shown that large deviations from these ideal equations can occur over the full range of composition (Foster and Woods 1971). A rigorous consideration of diffusion theory indicates that under these circumstances the diffusion coefficient becomes dependent on solute concentration, i.e. on the fraction of GaP in this case (see Tuck 1974). The dependence is not simple, since it is determined by the precise manner in which the activity of the GaP varies in the concentration range 0–100 percent. An anomalous diffusion profile might therefore be expected.

The depths of the phosphorus profiles were considerably greater than those for arsenic self-diffusion taken under similar conditions, possibly due to the fact that phosphorus atoms are smaller than arsenic. The penetrations were, nevertheless, sufficiently small for Jain *et al* to assume that the diffusion proceeded by a vacancy mechanism. The simplest such mechanism is the one proposed above for self-diffusion in GaAs, namely migration on the group V sub-lattice by exchange of a phosphorus atom and a vacancy. They also put forward a further rather intriguing possibility involving the movement of atoms round hexagonal rings in the (111) planes. The mechanism is illustrated in figure 7.3, and could apply equally well to self-diffusion of either group III or group V atoms; as drawn, it refers to group V. The ring shown at the centre of the unit cell contains three group III atoms and three group V sites. One of the group V sites is occupied by a vacancy. Five moves by the vacancy round the hexagon moves all atoms by one unit; unfortunately, this means that at this stage they are all on the wrong sites, i.e. five anti-site defects have been created. A further five steps moves all atoms onto correct sites and a given group V atom will then have moved to a next-nearest site. This method of migration has the advantage that the interactions are all between nearest-neighbour sites. On the other hand, it creates large numbers of anti-site defects and this would be expected to make it energetically unfavourable. Improvements have been suggested to make the mechanism more reasonable and these will be discussed later in the chapter.

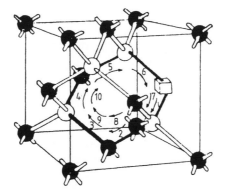

Figure 7.3 Ring mechanism, showing 10 jumps of vacancy (from Jain *et al* 1976). Reproduced by permission of *Am. Inst. Phys.*

7.2 InP and InAs

Despite the importance of InP as a material for opto-electronic devices, there appears to have been no recent self-diffusion study. Goldstein (1961) studied diffusion of both elements using techniques analogous to those employed in his GaAs work. Radioactive indium was plated onto InP slices and diffusions were performed in the temperature range 840–990 °C. The profiles were of standard form, as would be expected, and the diffusion coefficients followed an expression of the form of equation (7.1), with pre-exponential term $1 \times 10^5 \, cm^2 \, s^{-1}$ and activation energy $3.85 \pm 0.04 \, eV$. Phosphorus diffusions were from the vapour phase and carried out in the range 904–1010 °C. For these experiments, the pre-exponential term was $7 \times 10^{10}\ \mathrm{cm^2\,s^{-1}}$ and the activation energy $5.65 \pm 0.06\ \mathrm{eV}$.

Goldstein noted the fairly large difference in the two activation energies and, using the argument given in section 7.1.2, took this to indicate that both elements diffuse on their own sub-lattices. Once again, the value of diffusion coefficient at a given temperature proved to be smaller for the group V element. This seems to indicate that atomic site is of more significance than atomic size in determining D; the covalent tetrahedral radius of indium is 1.44 Å, while that of phosphorus is only 1.10 Å. Goldstein pointed out that if the diffusing atoms are neutral then the motion of a group V atom requires the re-ordering of five binding electrons, while motion of a group III needs only three. This could lead to the former atom requiring more energy to promote migration.

Self-diffusion of both indium and arsenic was studied by Kato *et al* (1969) and the results were broadly analogous to the results for GaAs and InP, with the indium coefficients greater than the arsenic. Diffusions were carried out from the vapour phase using radioactive sources. The results are rather less informative than most of the work described so far, since the diffusion profiles are given in terms of specific activity rather than absolute atomic concentration. The profiles were taken over the temperature range 740–900 °C and proved to be erfcs. Arrhenius plots gave pre-exponential values of $6 \times 10^5\ \mathrm{cm^2\,s^{-1}}$ and $3 \times 10^7\ \mathrm{cm^2\,s^{-1}}$ for indium and arsenic respectively and activation energies of 4.0 eV and 4.5 eV.

Arseni (1968) diffused radioactive phosphorus into InAs between 650 °C and 900 °C. Phosphorus pressures in the closed ampoules were 3–8 atm. The results differed in one respect from the P/GaAs work described in section 7.1.3; it was reported that all experiments gave rise to erfc profiles. The diffusion coefficients followed the relation

$$D = 126\ \exp\!\left(\frac{-2.7 \pm 0.2\ \mathrm{eV}}{kT}\right)\ \mathrm{cm^2\,s^{-1}} \qquad (7.3)$$

and, in common with the GaAs system, it was found that the values of the coefficient for phosphorus were very much greater than the self-diffusion coefficients of arsenic, the difference amounting to two to three orders of magnitude.

7.3 Self-Diffusion in InSb

The first measurements of self-diffusion in InSb were made by Boltaks and Kulikov (1957). Values of diffusion coefficient were obtained which were considerably greater than those reported by later workers, and it is likely that the results were affected by the presence of grain boundaries in the semiconductor crystals used. Eisen and Birchenall (1957) carried out radiotracer work and plotted diffusion profiles. Their curves showed two branches: a shallow component near the surface and a deeply-penetrating portion in the bulk. While profiles of this type are not unusual in impurity diffusion, it is surprising to see them occurring in self-diffusion experiments, which are essentially of the 'isoconcentration' type. The result was subsequently confirmed by Kendall and Huggins (1969), however, who showed fairly convincingly that the deeper-penetrating part of the profile could be attributed to surface pits which developed during the diffusion process. They therefore used the initial shallow portion of the profiles in calculating values of diffusion coefficient.

Figure 7.4 shows the two coefficients as functions of temperature. Both follow the expected Arrhenius form, with D_0 equal to $1.8 \times 10^{13} \, cm^2 s^{-1}$ and $3.1 \times 10^{13} \, cm^2 s^{-1}$ for indium and antimony respectively, and an activation energy of 4.3 eV for both elements. Diffusions were also carried out in which the ambient vapour pressure of antimony was a variable. It was found that changes in the group V over-pressure made no appreciable difference to the values of D. Kendall and Huggins concluded from this that in neither case could a simple vacancy mechanism be operative; such a mechanism gives a value of D which depends on the concentrations of vacancies and these, in turn, depend on the ambient vapour pressures, as outlined in Chapter 2. The concentration of divacancies $[V_{In}V_{Sb}]$, on the other hand, is independent of the ambient vapour pressures and it was suggested that diffusion of both elements proceeds by migration of this defect. The fact that both processes exhibited the same activation energy lends weight to the proposal. An approximate treatment was applied to describe the mechanism, and it was suggested that the experimental activation energy could be divided into component parts of 3.2 eV for the formation of a divacancy and 1.1 eV for the enthalpy of motion of the defect.

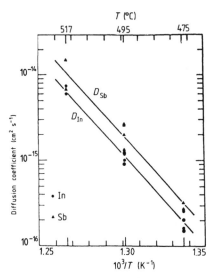

Figure 7.4 Self-diffusion coefficients in InSb (from Kendall and Huggins 1969).

7.4 Diffusion in GaSb

Diffusion of antimony was reported by Boltaks and Gutorov (1960), while diffusion of both elements was studied by Eisen and Birchenall (1957). Both groups used radiotracer techniques and material which was 'single crystal or slightly polycrystalline'. The work of the latter group was very similar to their InSb study reported in the previous section. A tail was found in the profiles, probably arising from the same pitting phenomenon found in InSb. No profiles are given in the Russian paper, so it is not clear whether their curves also demonstrated the tail effect. Certainly the results of the two groups are quite different. Both indicate that the diffusion coefficients follow the standard Arrhenius form. Eisen and Birchenall give values of D_0 of $3.2 \times 10^3 \, \mathrm{cm^2 \, s^{-1}}$ and $3.4 \times 10^4 \, \mathrm{cm^2 \, s^{-1}}$ for gallium and antimony respectively and activation energies of 3.15 eV and 3.45 eV. Boltaks and Gutorov give $8.7 \times 10^{-3} \, \mathrm{cm^2 \, s^{-1}}$ and 1.13 eV for antimony in GaSb. These latter results indicate very much higher values of D. The Russian workers also diffused indium into GaSb and again obtained high values of D; their diffusion coefficients exceeded those produced in the work of Mathiot and Edelin (1980), described below, by some four orders of magnitude. It seems likely that their results were affected to a significant extent by surface effects and by the poor quality of the GaSb used, which may well have had a high dislocation content.

The work on the diffusion of indium in GaSb by Mathiot and Edelin (1980) was rather more extensive than any of the previous studies. Experiments were carried out from 520 °C to the melting point of 720 °C, and diffusion profiles were obtained using SIMS. These workers paid rather more attention than is usual to the equilibrium conditions inside the diffusion ampoules. Before a diffusion run, the GaSb slice was, in most cases, heated at the diffusion temperature in a closed ampoule with a Ga–Sb source. The source was either on the gallium-rich side or the antimony-rich side of the phase diagram, corresponding either to point A or to point C on figure 7.5. During this anneal, the source would have been a mixture of solid and liquid phases, the compositions being given by the appropriate points on the liquidus and solidus curves at the anneal temperature. The object of this pre-diffusion saturation was to establish in the slice a stoichiometry which was the same as the solid phase in the source, i.e. a composition corresponding either to point B or to point D on the diagram. The diffusion treatments were carried out with the same experimental

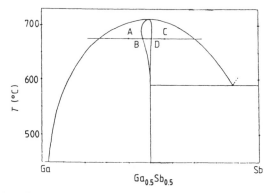

Figure 7.5 Phase diagram of GaSb. The width of the solid phase is greatly exaggerated (from Mathiot and Edelin 1980). Reproduced by permission of *Taylor and Francis Ltd*.

arrangement, except that a few percent of indium was added to the saturation source so that it became a diffusion source. Thus two sets of diffusion results were obtained: one for gallium-rich specimens and one for antimony-rich. At the melting point this distinction disappears, of course.

In common with other workers on InSb and GaSb, they found that a tail appeared on the experimental profiles at low concentrations. If this tail was ignored, the profiles were good erfc curves, allowing an unambiguous determination of D. A clear correlation was established between the concentration at which the tail became apparent and the dislocation content of the slice. A slice of dislocation density 6.5×10^4 cm^{-2} diffused at 660 °C, for instance, showed a good erfc profile extending from the surface and covering two orders of magnitude before any tail could be seen (see figure 7.6(a)). A sample with a density of 10^6 cm^{-2}, on the other hand, showed a diffusion profile completely corrupted by the tail (figure 7.6(c)).

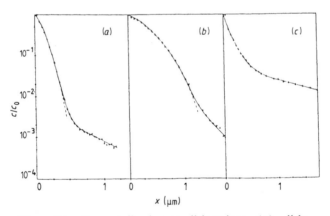

Figure 7.6 Curve tails due to dislocations. (a), dislocation density, $\rho = 6.5 \times 10^4$ cm^{-2}, Ga-rich sample; (b), $\rho = 3.5 \times 10^4$ cm^{-2}, Sb-rich sample; (c), $\rho = 10^6$ cm^{-2}, Ga-rich sample (from Mathiot and Edelin 1980). Reproduced by permission of *Taylor and Francis Ltd*.

The values of diffusion coefficient are shown in figure 7.7 as a function of inverted temperature. The antimony-rich results extend only to 590 °C because a eutectic point appears on the phase diagram at this temperature and a change in experimental tech-

nique would have been required to go below it. Note that the two
sets of results do not fall on a straight line; indeed this is hardly
possible since the two lines must curve round to meet at the melting
point. The reason is that the experiments were carried out at the
two extremes of permitted stoichiometry at each temperature, so in
both the gallium-rich and the antimony-rich cases the stoichiometry
was different at each temperature employed (this is clear from the
shape of the solid phase in figure 7.5). In principle, a much larger
set of experiments could have been carried out, using all values of
stoichiometry between the extremes. If this were done, the space
between the two lines on figure 7.7 would become filled with data
points. This is the sort of subtlety which is not often addressed in

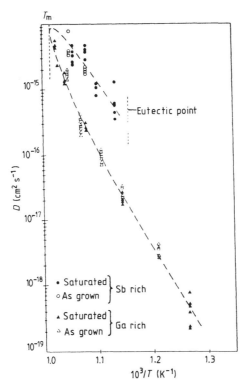

Figure 7.7 Temperature dependence of the diffusion
coefficient along the solidus curve of GaSb (from Mathiot
and Edelin 1980). Reproduced by permission of *Taylor and
Francis Ltd.*

diffusion papers because, in general, the work is not carried out with the rigour shown in this particular investigation. The extraction of activation energies from this sort of data is often done much too casually in compound semiconductor work.

The one unambiguous result one can take from figure 7.7 is that indium diffuses faster in antimony-rich than in gallium-rich material. A major difference between the two types, of course, is that the gallium vacancy concentration is greater in antimony-rich GaSb. The very low values of D which were found, together with the fact that indium occupies the gallium site, led the authors to propose that indium diffuses in GaSb by a simple vacancy mechanism on the group III sub-lattice. Such a mechanism would be expected to give a diffusion coefficient proportional to the gallium vacancy concentration in the crystal, C_v say. This conclusion can be tested by comparing the variation of D with temperature, shown in figure 7.7, with a similar plot giving the variation of C_v, shown in figure 7.8. This latter graph is taken from another paper by the same authors (Edelin and Mathiot 1980) in which they calculated C_v from published electrical results, using the assumption that the gallium vacancy acts as an acceptor in the semiconductor. Comparison of the two figures indicates a very similar variation, giving qualitative support to the proposed diffusion mechanism. Unfortunately, an attempt to use the data of figure 7.8 in a more quantitative way to derive the D v. $1/T$ curve was not quite so successful. The calculated curve did not agree in detail with figure 7.7. By way of explanation, it was suggested that the original calculation of C_v had underestimated the difference in vacancy concentration between gallium-rich and antimony-rich material.

A rather different interpretation of figure 7.7 was offered by Shaw (1983). He noted that the antimony vapour source during the diffusions would have been expected to be Sb_4, with vapour pressure P_{Sb_4}. To simplify the nomenclature, he put $(P_{Sb_4})^{1/4} = \theta$. Using the experimental work of Prasad *et al* (1979) and Gratton and Woolley (1980), he calculated θ for each of the In/GaSb diffusions and found that the curve of figure 7.7 fitted well to the function

$$D = 0.831 \exp\left(\frac{-2.87\,\text{eV}}{kT}\right)$$
$$+ 1.09 \times 10^{-2} \exp\left(\frac{-2.15\,\text{eV}}{kT}\right)\theta^2 \text{ cm}^2\,\text{s}^{-1} \quad (7.4)$$

i.e. two terms, one independent of θ (and thus of ambient antimony vapour pressure) and one proportional to θ^2. This was interpreted as indicating that diffusion proceeds by two different diffusion mechanisms, one represented by each term.

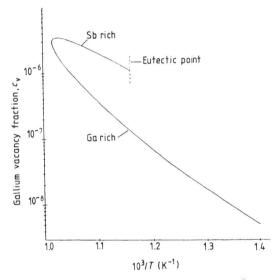

Figure 7.8 Calculated temperature dependence of the vacancy concentration (from Edelin and Mathiot 1980). Reproduced by permission of *Taylor and Francis Ltd*.

The simplest defect which could give rise to the first term is the divacancy ($V_{Ga}V_{Sb}$). The concentration of this vacancy pair is constant at constant temperature; this has been shown for the analogous case of GaP in the argument leading to equation (3.28). It was pointed out by Shaw that diffusion via a mixed divacancy had previously been proposed by Kendall and Huggins (1969) to explain their results on the self-diffusion of In in InSb (see section 7.3). The simplest defect having a concentration varying as θ^2 is ($V_{Ga}V_{Ga}$). This is essentially the same mechanism as that considered in Chapter 3 for the diffusion of sulphur in GaAs, and the required result is given in equation (3.5).

While it is certainly the case that equation (7.4) fits the results rather better than does the original calculation of Mathiot and Edelin (1980), it may be a little premature to assume that indium diffusion in GaSb proceeds by this fairly complicated parallel process rather than by the simpler vacancy mechanism suggested by the earlier workers. It is probably true to say that the more parallel mechanisms that are proposed, the more accurately can theoretical calculations be made to fit experimental results. More work is required before we will be able to decide between these two alternative explanations.

7.5   Diffusion in Superlattices

It was noted in Chapter 1 that major advances have been made in recent years in the development of techniques for the growth of single crystal semiconductor films. The principle feature of the films is that they can be made very thin indeed; thicknesses of 10 Å or so are now achievable. Two techniques are used for film preparation: metal-organic chemical vapour deposition (MOCVD) and molecular beam epitaxy (MBE). Both are able to mix layers of different semiconductors so that relatively thick 'sandwich' structures can be composed of, for instance, alternate layers of AlAs and GaAs. Such structures are often called 'superlattices' and they have important applications in the field of semiconductor devices. The AlAs–GaAs system is particularly convenient since the two semiconductors have almost identical lattice constants, so there is no mismatch between the layers. The solid solution $Al_xGa_{1-x}As$ exists for $0 < x < 1$ and matches to either parent. In other systems the two lattice constants do not match so well; in general it is still possible to grow the sandwich structures, but strain occurs at the interfaces. Such structures are called 'strained layer superlattices' (SLS).

7.5.1   Interdiffusion in superlattices

Consider a structure of 100 layers, alternately of AlAs and GaAs. If the sample is raised to diffusion temperature, interdiffusion would be expected to occur so that eventually it would become a homogeneous sample of $Al_xGa_{1-x}As$. The redistribution would be on the group III site and it seems reasonable to assume that the interdiffusion mechanism would be similar to the self-diffusion of aluminium in AlAs and gallium in GaAs. The experiment was carried out by Chang and Koma (1976) who used MBE to grow layers of thickness in the region of 1000 Å and carried out heat treatments in the range 850–1100 °C. Interdiffusion was monitored using Auger Electron Spectroscopy (AES). In this method, the semiconductor is bombarded with electrons, causing electrons to be emitted from the atoms at the surface. Some of the emitted electrons have energies characteristic of the atoms concerned, so the technique permits a chemical analysis of the surface (see, for example, Prutton 1983). It is possible to take AES data as a surface is sputtered, so a profile of the element under study can be obtained.

The profiles for a sample with layers of thickness 1550 Å, annealed at 992 °C for 14 hours, are shown in figure 7.9(a). The ratios of the Auger signals of Ga/As and Al/As are given, plotted against sputtering time, although the arsenic signal was essentially constant, as would be expected. After a calibration procedure to convert sputtering time to depth, a sequence of interdiffusion profiles can be obtained for increasing diffusion times; this is given in figure 7.9(b). Chang and Koma pointed out that what was actually being measured was interdiffusion of gallium and aluminium on a $Al_xGa_{1-x}As$ lattice. They came to the conclusion that D is a function of x, increasing as x decreases. At 1100 °C, the range of D for $0 < x < 1$ was 10^{-14}–10^{-15} cm^2 s^{-1}. This compares well with the value for self-diffusion of gallium in GaAs of 10^{-14} cm^2 s^{-1} found by Palfrey *et al* (1981) at the same temperature. At 800 °C, D was in the range 10^{-17}–10^{-18} cm^2 s^{-1}. Similar experiments were carried out on much thinner layers by Fleming *et al* (1980).

A series of experiments performed on MOCVD material was reported by Schlesinger and Kuech (1986). Multilayer structures were grown consisting of alternate layers of GaAs and $Al_{0.3}Ga_{0.7}As$. The former had thickness in the range 20–150 Å, while the cladding layers of $Al_{0.3}Ga_{0.7}As$ were 500 Å. The slices were subjected to heat treatments over the temperature range 650–910 °C for times between one and six hours. The characteristic photoluminescence from the GaAs wells, excited by a He–Ne laser, was studied at 77 K. After a heat treatment, the wavelength of the luminescence was found to decrease, and it was assumed that this was due to aluminium diffusing into the wells. Some fairly tortuous calculations permitted the increase in the well width to be estimated

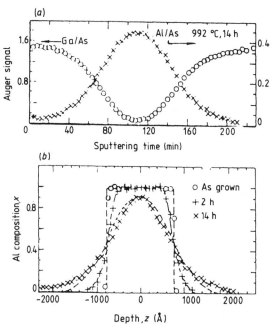

Figure 7.9 Profiles of interdiffusion between GaAs and AlAs at 992 °C. (*a*), Auger signals plotted against sputtering time; (*b*), Al composition against depth. Also given in (*b*), by broken curves, are the original square profile and a calculated profile for $D = 2 \times 10^{16} \, \text{cm}^2 \, \text{s}^{-1}$ (from Chang and Koma 1976). Reproduced by permission of *Am. Inst. Phys.*

from this wavelength change and an interdiffusion coefficient was found for each annealing temperature. The final result is shown in figure 7.10; it agrees quite well with the data of Chang and Koma (1976), but no mention is made in the paper of the possibility of D being a function of x. Note that the figure only covers the temperature range 820–910 °C; any changes in photoluminescence wavelength which may have occurred after anneals below this range were too small to be detected by the technique employed. An activation energy of about 6 eV can be deduced from the diagram.

There is very little published work on interdiffusion in strained superlattices. Kakimoto *et al* (1985) used Raman spectroscopy to monitor the loss of the layer structure by interdiffusion in

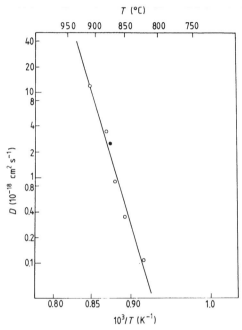

Figure 7.10 Interdiffusion coefficient of aluminium and gallium as a function of temperature (from Schlesinger and Kuech 1986). Reproduced by permission of *Am. Inst. Phys.*

GaAs/InAs SLS and compared the results with some taken from unstrained GaAs/AlAs structures. Both types of specimen were prepared by MBE, with individual layer thicknesses between 10 Å and 1000 Å. As might be expected, it was found that the strain-free lattices were more stable than the SLS. In the case of GaAs/AlAs, it was found that samples of smaller periodicity were less stable than those with larger periodicity; this is an interesting result which suggests that the interface between the two materials has a marked effect despite the almost perfect fit of the two lattices. Another interesting result was that for both types of specimen, superlattices grown at higher temperatures were less stable. Joncour *et al* (1985) annealed $GaAs/In_{0.14}Ga_{0.86}As$ SLS at $850\,^\circ C$. Again the samples were prepared by MBE; ten periods were used with thicknesses of 200 Å and 100 Å respectively for the GaAs and $In_{0.14}Ga_{0.86}As$ layers. An X-ray diffraction technique was used to detect variations in composition due to heat treatments. An estimated value of 10^{-18} cm^{2}s^{-1} was given for the interdiffusion coefficient at 850 °C.

7.5.2   *Diffusion-induced disorder (DID)*

Studies of the diffusion of zinc into superlattices yielded a quite remarkable result. It was found that the presence of the diffusing zinc increased the interdiffusion between AlAs and GaAs, for instance, by several orders of magnitude (Laidig *et al* 1981, 1982, 1983). This was demonstrated in elegant fashion by masking the whole surface of a sample except for a 10 μm strip through which the zinc could enter the material. After a ten minute anneal at 575 °C, the region under the strip took on a quite different form to the rest of the sample. Under the mask the layers remained distinct, while directly under the strip, where zinc diffusion had occurred, the sample became homogeneous Al$_x$Ga$_{1-x}$As. A photomicrograph demonstrating the effect is given in figure 7.11. Note that a magnification of 145 has been achieved in the diffusion direction (vertical in the figure) by angle-lapping the surface at an angle of about 0.4 °.

Lee and Laidig (1984) investigated the effect of varying the layer thickness. AlAs/GaAs superlattices were produced by MBE with periods 320 Å, 630 Å and 1100 Å (a period is the combined width of one AlAs layer and one GaAs layer). In all cases the total superlattice thickness was 1.8 μm. Zinc diffusion was carried out at 550 °C, 575 °C and 600 °C. The disorder was determined using AES and the zinc diffusion profile was plotted by SIMS. It was convenient to grade the degree of disorder into three categories: 'total', 'intermediate' and 'slight', and the result of this division is shown in figure 7.12 for the three types of specimen at the three temperatures employed. As might have been expected, the disorder is greater at greater diffusion temperatures. It is interesting, however, that the specimens of smaller period showed a greater degree of disorder at the end of a diffusion, suggesting once again that the layer interfaces play an important role in the effect. Some rough-and-ready techniques were used to extract approximate values of interdiffusion coefficient from the AES data. Coefficients in the region of 10^{-16} cm^{2}s^{-1} were obtained in the temperature range employed; the values were a little larger in the specimens of smaller period. Similarly, diffusion coefficients for the zinc were calculated

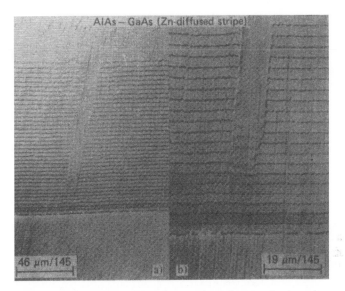

Figure 7.11 Shallow-angle cross section of a superlattice selectively diffused with zinc at 575 °C (from Laidig *et al* 1981). Reproduced by permission of *Am. Inst. Phys.*

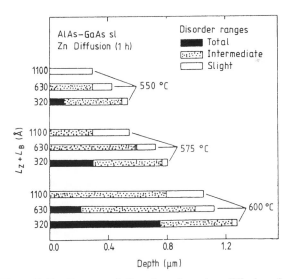

Figure 7.12 Degree of disorder after zinc diffusion for one hour at three temperatures. L_Z, L_B are, respectively, the thicknesses of GaAs and AlAs layers (Lee and Laidig 1984). Reproduced by permission of the *The Metallurgical Society*.

from the SIMS results. Again there was evidence to suggest that the diffusion was slightly greater in the finer structures. Comparison of the two sets of experimental data indicated that disordering of the superlattice sets in when the zinc concentration rises to about 10^{18} cm^{-3}. This last result was subsequently confirmed by Ishida *et al* (1985).

An interesting variation on the above experiment was carried out by Camras *et al* (1983). They prepared MBE specimens of alternate layers of GaP and GaAs$_{0.4}$P$_{0.6}$, giving a strained-layer structure. The samples all consisted of 40 periods of layer width 120 Å. Zinc was diffused through a narrow opening in a Si$_3$N$_4$ cap for 14 hours at 825 °C. Mixing occurred under the strips, but not elsewhere, just as in the AlAs/GaAs samples described above. This is indeed an unexpected result; zinc occupies the group III site in the III–V compounds and it is perhaps not too surprising to find its diffusion interacting in some way with self diffusion of group III atoms. It is not nearly so easy to see why it should interfere with self-diffusion of group V atoms, however.

Matters were further complicated by a result from Wu *et al* (1986) which suggested that the disordering effect does not occur if the superlattice layers are doped with certain atoms. They prepared structures consisting of 100 Å layers of beryllium-doped GaAs alternating with 90 Å layers of tin-doped Al$_{0.3}$Ga$_{0.7}$As. The doping level in each case was about 2×10^{17} cm^{-3}. Zinc was diffused into the superlattice for 30 minutes at 600 °C. Observation under an electron microscope revealed that the layers remained quite distinct after the diffusion.

A number of groups have reported that silicon acts in much the same way as zinc in initiating DID. Meehan *et al* (1984) diffused silicon into a superlattice structure consisting of 40 periods of 280 Å GaAs and 320 Å Al$_{0.6}$Ga$_{0.4}$As layers. Si$_3$N$_4$ was used to mask the surface, leaving 15 μm strips through which silicon could enter the sample. The diffusion was at 850 °C for ten hours. At the end of this time, the layers were mixed under the strip but unaffected elsewhere, as with the analogous zinc experiment.

Kawabe *et al* (1984) and Iwata *et al* (1985) performed a slightly different experiment in which the layers were uniformly doped with silicon during growth. After growth, the superlattice structure was perfect, but heat treatment for two hours at 850 °C induced mixing of the layers if the silicon concentration was greater than about

4×10^{18} cm^{-3}. This experiment established that there is no need for a concentration gradient to exist for the disordering effect to be observed. Silicon was also grown into the specimens of Gonzalez *et al* (1986) who produced a fairly complicated structure consisting of an AlAs/GaAs superlattice grown on a relatively thick silicon-doped layer of Al$_{0.3}$Ga$_{0.7}$As. In this case, silicon did diffuse during growth and disorder the superlattice. The growth temperature of 700 °C was rather higher than is normally used in MBE, however.

It has been mentioned above that there is evidence that beryllium can inhibit disorder induced by zinc diffusion. There is even stronger evidence that it inhibits the silicon-induced effect. Kawabe *et al* (1985) grew AlAs/GaAs superlattices using MBE with all layers 150 Å in thickness. They found that doping the structures with beryllium during growth and then heat-treating did not bring about mixing. Further structures were grown with undoped, silicon-doped and silicon-plus-beryllium-doped sections. These were then annealed at 780 °C for two hours. The layer structure was monitored after the anneal using AES profiling. It was found that the undoped section did not disorder and the silicon-doped section did, in agreement with the previous work. The section doped with both elements disordered if the beryllium doping was less than about 10^{19} cm^{-3}. For higher concentrations of beryllium, the mixing was suppressed. A similar result was obtained by Kobayashi *et al* (1986). They put forward a model for the inhibiting action based on the mechanism for silicon diffusion in GaAs proposed by Greiner and Gibbons (1984). According to this model, diffusion of silicon occurs via substitutional exchange of (Si$_{III}$Si$_V$) pairs with vacancies (see Chapter 3). It was suggested that beryllium, an acceptor occupying the group III site, forms donor–acceptor pairs with silicon of the form (Be$_{III}$Si$_{III}$). This prevents the formation of the silicon pairs and reduces normal silicon diffusion, thereby inhibiting the silicon-induced mixing effect.

At present there is conflicting evidence as to whether donors occupying group V sites in the superlattice can cause group III disordering. Rao *et al* (1985) diffused sulphur into Al$_x$Ga$_{1-x}$As/ GaAs structures at 850 °C from a vapour source and reported mixing of the lattices. The disorder was detected by photoluminescence; a reduction in luminescence wavelength after diffusion was taken to indicate mixing. The same method has been employed by a number of workers but, in general, it has been used in combination with other techniques, such as AES, TEM etc. This is probably a wise policy, since interpretation of photoluminescence data is not always straight-forward. Dobisz *et al* (1986) ion-implanted both silicon and sulphur into AlAs/GaAs superlattices and then heat-treated the samples for two hours at 800 °C. DID was found in the silicon-implanted specimens, but not in those containing sulphur. On the other hand, Gavrilovic *et al* (1985) implanted a wide range of atoms, including sulphur, into the same structure and reported DID in all specimens after annealing. The problem with using ion implantation rather than simple diffusion to study DID, of course, is that it is difficult to separate diffusion effects from those due to damage.

7.5.3   Models for DID

Laidig *et al* (1982) attempted to explain their results on zinc-induced disordering using a modification of the substitutional–interstitial mechanism for the diffusing zinc atoms. They suggested that in the transfer of a zinc atom from an interstitial to a substitutional site an intermediate stage could occur, consisting of a complex formed by an interstitial zinc atom and a group III vacancy:

$$Zn_i^+ + V_{III} \rightleftharpoons (Zn_i V_{III})^+ \rightleftharpoons Zn_{\overline{III}} + 2h \qquad (7.5)$$

where V_{III} and $Zn_{\overline{III}}$ represent a vacancy and a zinc atom on a group III site (i.e. either an aluminium or a gallium site) and h is a hole. It was suggested that the vacancy part of the complex could take part in the self-diffusion process i.e. that the presence of the zinc could effectively increase the concentration of group III vacancies and hence aid self-diffusion.

A rather more detailed approach to the mechanism of DID was taken by Van Vechten (1982, 1984). He showed that the diffusion process proposed by Jain *et al* (1976) to explain the diffusion of phosphorus in GaAs (see figure 7.3) could be adapted to account for disordering due to both silicon and zinc. The mechanisms involve vacancies and anti-site defects and it was assumed that all of these defects are charged. Vacancies on group III and V sites were given single negative and single positive charges respectively. Anti-site defects were assumed doubly charged, with a group III atom on a group V site negative, and a group V atom on a group III site positive. Atoms on the correct sites are neutral. The mechanism for silicon diffusion is shown in figure 7.13. The figure shows adjacent 'hexagons' on the (111) plane of the zincblende structure. The only difference between this mechanism and the one originally described by Jain *et al* is that one of the atoms in a hexagon is silicon. As noted earlier, this atom sits quite happily on either site, acting either as the donor Si_{III}^+ or acceptor Si_V^-. Once again, ten jumps of the vacancy are needed to rotate the atoms to next-nearest-neighbour positions, and anti-site defects are created on the

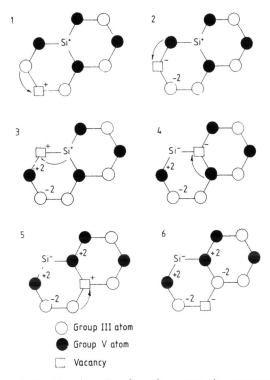

○ Group III atom
● Group V atom
☐ Vacancy

Figure 7.13 Showing the first five steps in proposed mechanism for interaction of silicon diffusion and self-diffusion (Van Vechten 1984). A further five steps are needed to remove the anti-site defects. It is assumed that vacancies are charged negatively on group III sites and positively on group V sites. Anti-site defects are charged -2 on group III sites and $+2$ on group V sites.

way. Because of the presence of the silicon atom, however, a maximum of four such defects is created rather than the five shown by the simple Jain self-diffusion mechanism. Van Vechten pointed out that this should make the self-diffusion process easier and therefore increase the self-diffusion coefficient. It is not at all clear, however, that this effect would be large enough to give the observed experimental result. The proposed diffusion mechanism is also incompatible with that proposed for silicon diffusion in GaAs by Greiner and Gibbons (1984).

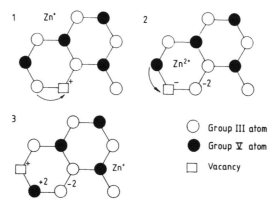

Figure 7.14 Showing interaction between zinc interstitial and charged defects as proposed by Van Vechten (1982). Note that it is assumed that the zinc atom can be either singly or doubly charged.

A variation on the same theme was proposed for zinc-induced disorder. Here the positively-charged zinc interstitial plays a role, interacting with the Jain mechanism. Again the vacancy moves anti-clockwise round a (111) hexagon, leaving in its wake a trail of charged anti-site defects (figure 7.14). If the defects give rise to a net negative charge, the interstitial is attracted towards them and reduces the potential energy of the configuration. If they make a net positive charge, the zinc atom moves away and the energy is unaffected. This movement of the atom can occur quite easily, since zinc is extremely mobile in the III–Vs when interstitial. A net reduction in energy is obtained over the ten steps and the effective self-diffusion coefficient is again increased.

It was demonstrated by Van Vechten (1982) that if a divacancy, rather than a single vacancy, were to be involved in the above process, then rather fewer anti-site defects would be produced. This version of the mechanism is shown in figure 7.15. The first move creates an anti-site defect which is removed at the second step. In the absence of the zinc interstitial, this mode of self-diffusion would move roughly equal numbers of group III and group V atoms. The Coulombic attraction between the zinc ion and the negatively-charged defect (group III atom on group V site), however, enhances the rate of vacancy diffusion via the mode that mixes atoms on the group III sub-lattice. Thus steps such as 1–3 in figure 7.15 are strongly favoured in the presence of zinc. The mechanism would therefore appear to predict mixing of the group III sub-lattice during zinc diffusion, but not the group V. Unfortunately, as noted above, DID has been reported in $GaAs_xP_{1-x}/GaP$ superlattices. Camras *et al* (1983), who reported these experiments, were of the opinion that they rendered the divacancy mechanism untenable.

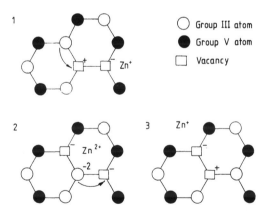

Figure 7.15 Showing divacancy version of mechanism of figure 7.14 (after Van Vechten 1982).

7.5.4 Uses of DID

Examples have been given earlier in the chapter of disorder being selectively introduced into a superlattice structure by partially masking the surface of a sample against diffusion. The region of superlattice under the masked area remains ordered, and the remainder becomes disordered. The procedure of producing strips of ordered material into a disordered matrix has considerable potential for use in opto-electronics. The difference in refractive index between ordered and disordered material, for instance, gives rise to the use of the strips as waveguides and as the active regions in quantum well lasers (Meehan *et al* 1984, Fukuzawa *et al* 1984, Deppe *et al* 1987). In the latter case, the electrical barrier between the ordered and disordered portions also acts to confine the carriers.

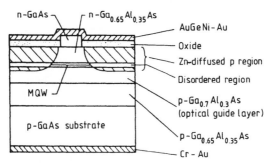

Figure 7.16 Multi-quantum-well laser fabricated using diffusion-induced disorder (from Fukuzawa *et al* 1984). Reproduced by permission of *Am. Inst. Phys.*

An example of the use of DID in forming the active region of a multi-quantum-well (MQW) laser is shown in figure 7.16. The layers, grown by MBE, were a 3 μm $Al_{0.35}Ga_{0.65}As$ cladding layer, doped with beryllium; a 1 μm $Al_{0.3}Ga_{0.7}As$ optical guide layer, also beryllium-doped; an undoped MQW active layer composed of five 80 Å GaAs wells separated by four 120 Å $Al_{0.3}Ga_{0.7}As$ barriers; a 1 μm $Al_{0.35}Ga_{0.65}As$ cladding layer (silicon doped) and a 1 μm GaAs cap layer. Zinc was diffused into this structure, leaving a 4 μm stripe unaffected by using a Si_3N_4 mask. The active MQW is simply and controllably defined by the edge of the zinc diffused region. Recently, Meehan *et al* (1986) have speculated on the possibility of creating three-dimensional structures in superlattices by creating some desired pattern using DID and then subjecting the sample to further layer growth, followed by more DID.

References

Allen G A 1968 *J. Phys. D: Appl. Phys.* **1** 593

Allen J W 1960 *J. Phys. Chem. Solids* **15** 134

Allison H W and Fuller C S 1965 *J. Appl. Phys.* **36** 2519

Ambridge T and Faktor M M 1975 *Gallium Arsenide and Related Compounds 1974* (Inst. Phys. Conf. Ser. 24) p 320

Ando H, Susa N and Kanbe H 1981 *Jap. J. Appl. Phys.* **20** L197

Antell G R 1965 *Solid State Electron.* **8** 943

Arnold N, Schmitt R and Heime K 1984 *J. Phys. D: Appl. Phys.* **17** 443

Arseni K A, Boltaks B I and Rembeza S I 1967 *Sov. Phys.–Solid State* **8** 2248

Arseni K A 1968 *Sov. Phys.–Semicond.* **2** 367

Arseni K A 1969 *Sov Phys.–Solid State* **10** 2263

Arseni K A and Boltaks B I 1969 *Sov. Phys.–Solid State* **10** 2190

Ball R K, Hutchinson P W and Dobson P S 1981 *Phil. Mag.* A **43** 1299

Black J F and Jungbluth E D 1967a *J. Electrochem. Soc.* **114** 181

Black J F and Jungbluth E D 1967b *J. Electrochem. Soc.* **114** 188

Blum S E, Small M B and Gupta D 1983 *Appl. Phys. Lett.* **42** 108

Bolkhovityanov Y B and Bolkhovityanova R I 1975 *Phys. Stat. Solidi* (a) **27** 673

Boltaks B I, Ksendzov S and Rembeza S I 1969 *Sov. Phys.–Solid State* **10** 2186

Boltaks B I, Kulikov G S, Nikulitsa I N and Shishiyanu F S 1975 *Inorg. Mater.* **11** 292

Boltaks B I, Rembeza S I and Bakhadyrkhanov M K 1968 *Sov. Phys.–Solid State* **10** 432

Boltaks B I, Rembeza S I and Sharma B L 1967 *Sov. Phys.–Solid State* **1** 196

Boltaks B I and Gutorov Y A 1960 *Sov. Phys.—Solid State* **1** 930

Boltaks B I and Kulikov G S 1957 *Sov. Phys.—Tech. Phys.* **2** 67

Boltaks B I and Shishiyanu F S 1964 *Sov. Phys.—Solid State* **5** 1680

Boltaks B I and Sokolov V I 1964 *Sov. Phys.—Solid State* **6** 600

Boltzmann L 1894 *Ann. Phys. Leipzig* **53** 959

Brozel M R, Foulkes E J and Tuck B 1982 *Phys. Stat. Solidi* (a) **72** K159

Brozel M R, Foulkes E J, Tuck B, Goswami N K and Whitehouse J E 1983 *J. Phys D:Appl. Phys.* **16** 1085

Brozel M R, Tuck B and Foulkes E J 1981 *Electron. Lett.* **17** 532

Brozel M R, Butler J, Newman R C, Ritson A, Stirland D J and Whitehead C 1978 *J. Phys. C: Solid State Phys.* **11** 1857

Buehler M G 1966 *Stanford Res. Rep.* EEL-66-064

Campbell A N, Wagemann R and Ferguson R B 1970 *Canad. J. Chem.* **48** 1703

Camras M D, Holonyak N, Hess K, Ludowise M J, Dietze W T and Lewis C R 1983 *Appl. Phys. Lett.* **42** 185

Carslaw H S and Jaeger J C 1959 *Conduction of Heat in Solids* (Oxford University Press)

Casey H C 1973 *Atomic Diffusion in Semiconductors* ed. D Shaw (London: Plenum) p 351

Casey H C and Panish M B 1968 *Trans. AIME* **242** 406

Chang L L and Casey H C 1964 *Solid State Electron.* **7** 481

Chang L L and Koma A 1976 *Appl. Phys. Lett.* **29** 138

Chang L L and Pearson G L 1964a *J. Appl. Phys.* **35** 374

Chang L L and Pearson G L 1964b *J. Appl. Phys.* **35** 1960

Chaoui R 1981 PhD Thesis, University of Nottingham, UK

Chaoui R and Tuck B 1983a *J. Phys. D: Appl. Phys.* **16** 1721

Chaoui R and Tuck B 1983b *Phys. Stat. Solidi* (a) **77** K189

Chevrier J, Armand M, Huber A-M and Linh N T 1980 *J. Electron. Mater.* **9** 745

Chiang S Y and Pearson G L 1975 *J. Appl. Phys.* **46** 2986

Chin A K, Camlibel I, Sheng T T and Bonner W A 1983b *Appl. Phys. Lett.* **43** 495

Chin A K, Camlibel I, Dutt B V, Swaminathan V, Bonner W A and Ballman A A 1983a *Appl. Phys. Lett.* **42** 901

Clegg J B 1982 *Semi-Insulating III–V Materials, Evian* ed. S Makram-Ebeid and B Tuck (Nantwich: Shiva) p 80

Colliver D J, Irving L D, Pattison J E and Rees H D 1974 *Electron. Lett.* **10** 221

Crank J 1956 *Mathematics of Diffusion* (Oxford University Press)

Crank J and Nicholson P 1947 *Proc. Camb. Phil. Soc.* **43** 50

Cunnell F A and Gooch C H 1960 *J. Phys. Chem. Solids* **15** 127

Da Cunha S F and Bougnot J 1974 *Phys. Stat. Solidi(a)* **22** 205

Davies D E, Potter W D and Lorenzo J P 1978 *J. Electrochem. Soc.* **125** 1845

Deal M D, Gasser R A and Stevenson D A 1985 *J. Phys. Chem. Solids* **46** 859

Deal M D and Stevenson D A 1984 *J. Electrochem. Soc.* **131** 2343

Deal M D and Stevenson D A 1986 *J. Appl. Phys.* **59** 2398

Debney B T and Jay P R 1980 *Semi-Insulating III–V Materials, Nottingham* ed. G J Rees (Nantwich: Shiva) p 305

Deppe D G, Jackson G S, Holonyak N, Burnham R D and Thornton R L 1987 *Appl. Phys. Lett.* **50** 632

Dlubek G, Brummer O, Plazaola F, Hautojarvi P and Naukkarinen L 1985 *Appl. Phys. Lett.* **46** 1136

Dobisz E A, Tell B, Craighead H G and Tomargo M C 1986 *J. Appl. Phys.* **60** 4150

Dutt B V, Chin A K, Camlibel I and Bonner W A 1984 *J. Appl. Phys.* **56** 1630

Dutt B V, Chin A K and Bonner W A 1981 *J. Electrochem. Soc.* **128** 2014

Dutt B V and Brasen D 1983 *J. Electrochem. Soc.* **130** 207

Edelin G and Mathiot D 1980 *Phil. Mag.* B **42** 95

Edmond J T 1960 *J. Appl. Phys.* **31** 1428

Eisen F H and Birchenall C E 1957 *Acta Metall.* **5** 265

Eu V, Feng M, Henderson W B, Kim H B and Whelan J M 1980 *Appl. Phys. Lett.* **37** 473

Fane R W and Goss A J 1963 *Solid State Electron.* **6** 383

Farges R W, Jacob G, Schemali C, Martin G M, Mircea-Roussel A and Hallais J 1982 *Semi-Insulating III–V Materials, Evian* ed. S Makram-Ebeid and B Tuck (Nantwich: Shiva)

Fleming R M, McWhan D B, Gossard A C, Wiegmann W and Logan R A 1980 *J. Appl. Phys.* **51** 357

Foster L M and Woods J F 1971 *J. Electrochem. Soc.* **118** 1175

Frank F C and Turnbull D 1956 *Phys. Rev.* **104** 617

Frieser R G 1965 *J. Electrochem. Soc.* **112** 697

Fujiwara Y, Kojima A, Nishino T and Hamakawa Y 1985 *Jap. J. Appl. Phys.* **24** 1479

Fukuzawa T, Semura S, Saito H, Ohta T, Uchida Y and Nakashima H 1984 *Appl. Phys. Lett.* **45** 1

Fuller C S, Allison H W and Wolfstirn K B 1964 *J. Phys. Chem. Solids* **25** 1329

Fuller C S and Wolfstirn K B 1967 *J. Electrochem. Soc.* **114** 856

Gansauge P and Hoffmeister W 1966 *Solid State Electron.* **9** 89

Gavrilovic P, Deppe D G, Meehan K, Holonyak N, Coleman J J and Burnham R D 1985 *Appl. Phys. Lett.* **47** 130

Gibbon C F and Ketchow D R 1971 *J. Electrochem. Soc.* **118** 975

Goldstein B 1961 *Phys. Rev.* **121** 1305

Goldstein B and Dobin C 1962 *Solid State Electron.* **5** 411

Goldstein B and Keller H 1961 *J. Appl. Phys.* **32** 1180

Gonzalez L, Clegg J B, Hilton D, Gowers J P, Foxon C T and Joyce B A 1986 *Appl. Phys.* A **41** 237

Gosele U and Morehead F 1981 *J. Appl. Phys.* **52** 4617

Gratton M F and Woolley J C 1980 *J. Electrochem. Soc.* **127** 55

Gray K W, Pattison J E, Rees H D, Prew B A, Clarke R C and Irving L D 1975 *Proc. Fifth Biennial Cornell Electrical Engineering Conf.* (Ithaca, NY: Cornell UP) p 125

Greiner M E and Gibbons J F 1985 *J. Appl. Phys.,* **57** 5181

Greiner M E and Gibbons J F 1984 *Appl. Phys. Lett.* **44** 750

Hall R N and Racette J H 1964 *J. Appl. Phys.* **35** 379

Hallais J, Mircea-Roussel A, Farges J P and Poiblaud G 1977 *Gallium Arsenide and Related Compounds 1976* (Inst. Phys. Conf. Ser. 33b) p 220

Hammer R 1969 *Rev. Sci. Instrum.* **40** 292

Hasegawa H and Hartnagel H 1975 *J. Electrochem. Soc.* **123** 713

Holmes D E, Wilson R G and Yu P W 1981 *J. Appl. Phys.* **52** 3396

Holmes D E, Elliot K R, Chen R T and Kirkpatrick C G 1982 *Semi-Insulating III–V Materials, Evian* ed. S Makram-Ebeid and B Tuck (Nantwich: Shiva) p 19

Huber A M, Morillot G, Linh N T, Favernnec P N, Deveaud B and Toulouse B 1979 *Appl. Phys. Lett.* **34** 858

Huber A M, Morillot G, Bonnet M, Merenda P and Bessoneau G 1982 *Appl. Phys. Lett.* **41** 638

Hutchinson P W and Ball R K 1982 *J. Mater. Sci.* **17** 406

Hwang C J 1968 *J. Appl. Phys.* **39** 5347

Ilegems M 1977 *J. Appl. Phys.* **48** 1278

Ilegems M and O'Mara W C 1972 *J. Appl. Phys.* **43** 1190

Ishida K, Ohta T, Semura S and Nakashima H 1985 *Jap. J. Appl. Phys.* **24** L620

Iwata N, Matsumoto Y and Baba T 1985 *Jap. J. Appl. Phys.* **24** L17

Jain G C, Sadana D F and Das B K 1976 *Solid State Electron.* **19** 731

Joncour M C, Charass M N and Burgeat J 1985 *J. Appl. Phys.* **58** 3373

Jordan A S 1971 *Metal Trans.* **2** 1965

Jordan A S 1982 *Semi-Insulating III–V Materials, Evian* ed. S Makram-Ebeid and B Tuck (Nantwich: Shiva) p 253

Kadhim M A and Tuck B 1972 *J. Mater. Sci.* **7** 68

Kakimoto K, Ohno H, Katsumi R, Abe Y, Hasegawa H and Katoda T 1985 *Gallium Arsenide and Related Compounds 1984* (Inst. Phys. Conf. Ser. 74) p 253

Karelina T A, Lavrishchev T T, Prikhod'ko G L and Khludkov S S 1974 *Inorganic Mater.* **10** 194

Kasahara J and Watanabe N 1982 *Semi-Insulating III–V Materials, Evian* ed. S Makram-Ebeid and B Tuck (Nantwich: Shiva) p 238

Kato H, Yokozawa M, Kohara R, Okabayashi Y and Takayanagi S 1969 *Solid State Electron.* **12** 137

Kawabe M, Shimizu N, Hasegawa F and Nannichi Y 1985 *Appl. Phys. Lett.* **46** 849

Kawabe M, Matsura N, Shimizu N, Hasegawa F and Nannichi Y 1984 *Jap. J. Appl. Phys.* **23** L623

Ke W K, Hamilton B, Peaker A R, Brozel M, Tuck B and Wight D R 1985 *Solid State Electron.* **28** 611

Kendall D L 1968 *Semiconductors and Semimetals* vol 4 ed. R K Willardson and A C Beer (New York: Academic Press) p 163

Kendall D L and Huggins R A 1969 *J. Appl. Phys.* **40** 2750

Khludkov S S, Prikhod'ko G L and Karelina T A 1972 *Inorg. Mater.* **8** 914

Khludkov S S and Lavrishchev T T 1976 *Inorg. Mater.* **12** 972

Kundukhov R M, Metreveli S G and Siukaev N V 1967 *Sov. Phys.–Semicond.* **1** 765

Klein P B, Nordquist P E R and Siebenmann P G 1980 *J. Appl. Phys.* **51** 4861

Klein T and Beale J R A 1966 *Solid State Electron.* **9** 59

Kobayashi J, Nakajima M, Fukunaga T, Takamori T, Ishida K, Nakashima H and Ishida K 1986 *Jap. J. Appl. Phys.* **25** L736

Kolodny A and Shappir J 1978 *J. Electrochem. Soc.* **125** 1530

Kressel H, Nelson H and Hawrylo F Z 1968 *J. Appl. Phys.* **39** 5647

Keubart W, Hildebrand O, Marten H W and Arnold N 1983 *Gallium Arsenide and Related Compounds 1982* (Inst. Phys. Conf. Ser. 65) p 597

Laidig W D, Holonyak N, Coleman J J and Dapkus P D 1982 *J. Electron. Mater.* **11** 1

Laidig W D, Holonyak N, Camras M D, Hess K, Coleman J J, Dapkus P D and Bardeen J 1981 *Appl. Phys. Lett.* **38** 776

Laidig W D, Lee J W, Chiang P K, Simpson L W and Bedair S M 1983 *J. Appl. Phys.* **54** 6382

Larrabee G B and Osborne J F 1966 *J. Electrochem. Soc.* **113** 564

Lee J W and Laidig W D 1984 *J. Electron. Mater.* **13** 147

Lehovic K and Slobodsky A 1961 *Solid State Electron.* **3** 45

Linh N T, Huber A M, Etienne P, Morillot G, Duchemin P and Bonnet M 1980 *Semi-Insulating III–V Materials, Nottingham* ed. G J Rees (Nantwich: Shiva) p 206

Logan R M and Hurle D T J 1979 *J. Phys. Chem. Solids* **32** 1739

Longini R L 1962 *Solid State Electron.* **5** 127

Lum W Y and Clawson A R 1979 *J. Appl. Phys.* **50** 5296

Luther L C and Wolfstirn K B 1973 *J. Electron. Mater.* **2** 375

Madelung O 1964 *Physics of III–V Compounds* (New York: Wiley) p 18

Magee T J, Peng J, Hong J D, Evans C A, Deline V R and Malbon R M 1979 *Appl. Phys. Lett.* **35** 277

Magee T J, Hong J, Deline V R and Evans C A 1980 *Appl. Phys. Lett.* **37** 53

Matano C 1932 *Jap. J. Phys.* **8** 109

Mathiot D and Edelin G 1980 *Phil. Mag.* A **41** 447

Matino H 1974 *Solid State Electron* **17**. 35

McLevige W V, Vaidyanathan K V, Streetman B G, Ilegems M, Comas J and Plew L 1978 *Appl. Phys. Lett.* **33** 127

Meehan K, Brown J M, Holonyak N, Burnham R D, Paoli T L and Streifer W 1984 *Appl. Phys. Lett.* **44** 700

Meehan K, Hsieh K C, Costrini G, Kaliski R W, Holonyak N and Coleman J J 1986 *Appl. Phys. Lett.* **48** 861

Meehan K, Holonyak N, Brown J M, Nixon M A and Gavrolic P 1984 *Appl. Phys. Lett.* **45** 549

Messham R L, Majerfeld A and Bachmann K J 1982 *Semi-Insulating III–V Materials, Evian* ed. S Makram-Ebeid and B Tuck (Nantwich: Shiva) p 75

Mircea-Roussel A, Jacob G and Hallais J P 1980 *Semi-Insulating III–V Materials, Nottingham* ed. G J Rees (Nantwich: Shiva) p 133

Mozzi R L and Lavine J M 1970 *J. Appl. Phys.* **41** 280

Nishizawa J, Sinozaki S and Ishida K 1973 *J. Appl. Phys.* **44** 1638

Nygren A S and Pearson G L 1969 *J. Electrochem. Soc.* **116** 648

Oberstar J D, Streetman B G, Baker J E and Williams P 1981 *J. Electrochem. Soc.* **128** 1814

Ohno H, Ushirokawa A and Katoda T 1979 *J. Appl. Phys.* **50** 8226

van Ommen A H 1983 *J. Appl. Phys.* **54** 5055

Orth R W and Watt L A K 1965 *J. Phys. Chem. Solids* **26** 197

Palfrey H D, Brown M and Willoughby A F W 1981 *J. Electrochem. Soc.* **128** 2224

Palfrey H D, Brown M and Willoughby A F W 1983 *J. Electron. Mater.* **12** 863

Panish M B 1966a *J. Phys. Chem. Solids* **27** 291

Panish M B 1966b *J. Electrochem. Soc.* **113** 224

Panish M B 1966c *J. Electrochem. Soc.* **113** 861

Panish M B 1967 *J. Electrochem Soc.* **114** 1161

Panish M B 1973 *J. Appl. Phys.* **44** 2659

Panish M B 1974 *J. Crystal Growth* **27** 6

van der Pauw L J 1958 *Phil. Res. Rep.* **13** 1

Petritz A 1958 *Phys. Rev.* **110** 1254

Potts H R and Pearson G L 1966 *J. Appl. Phys.* **37** 2098

Prasad R, Venugopal V, Singh Z and Sood D D 1979 *J. Chem. Thermodynamics* **11** 963

Prince F C, Oren M and Lam M 1986 *Appl. Phys. Lett.* **48** 546

Prutton M 1983 *Surface Physics* 2nd Edition (Oxford: Clarendon Press)

Rao E V K, Thibierge H, Brillouet F, Alexandre F and Azoulay R 1985 *Appl. Phys. Lett.* **46** 867

Rekalova G I, Kebe U and Mezrina L A 1971a *Sov. Phys.–Semicond.* **5** 685

Rekalova G I, Shakhov A A and Gavrushko A A 1969 *Sov. Phys.–Semicond.* **2** 1452

Rekalova G I, Kebe U, Persiyanov T V, Krymov V M and Krymova E D 1971b *Sov. Phys.–Semicond.* **5** 134

Rembeza S I 1967 *Sov. Phys.—Semicond.* **1** 516

Rembeza S I 1969 *Sov. Phys.—Semicond.* **3** 519

Rideout V L 1975 *Solid State Electron.* **18** 541

Rumsby D, Ware R M and Whittaker M 1980 *Semi-Insulating III–V Materials, Nottingham* ed. G J Rees (Nantwich: Shiva) p 59

Rybka V, Morihiro Y and Aoki M 1962 *J. Phys. Soc. Japan* **17** 1812

Schillmann E 1962 *Compound Semiconductors* vol 1 ed. R K Willardson and H L Goering (New York: Reinhold) p 358

Schlesinger T E and Kuech T 1986 *Appl. Phys. Lett.* **49** 519

Schneider M and Nebauer E 1975 *Phys. Stat. Solidi (a)* **32** 333

Schwuttke G H and Rupprecht H 1966 *J. Appl. Phys.* **37** 167

Seltzer M S 1965 *J. Phys. Chem. Solids* **26** 243

Sharma B L, Purohit R K and Mukerjee S N 1971 *J. Phys. Chem. Solids* **32** 1397

Shaw D 1974 *Atomic Diffusion in Semiconductors* ed. D Shaw (London: Plenum) p 1

Shaw D 1983 *J. Phys. C: Solid State Phys.* **16** L839

Shaw D 1984 *Phys. Stat. Solidi (a)* **86** 629

Shaw D and Showan S R 1969a *Phys. Stat. Solidi* **32** 109

Shaw D and Showan S R 1969b *Phys. Stat. Solidi* **34** 475

Shih K K, Allen J W and Pearson G L 1968a *J. Phys. Chem. Solids* **29** 367

Shih K K, Allen J W and Pearson G L 1968b *J. Phys. Chem. Solids* **29** 379

Shishiyanu F S, Gheorghiu V Gh and Palazov S K 1977 *Phys. Stat. Solidi* (a) **40** 29

Shishiyanu F S and Boltaks B I 1966 *Sov. Phys.—Solid State* **8** 1053

Showan S R and Shaw D 1969 *Phys. Stat. Solidi* **32** 97

Skolnick M S, Brozel M R and Tuck B 1982, *Solid-State Commun.* **43** 379

Skoryatina E A 1986 *Sov. Phys.—Semicond.* **20** 1177

Small M B, Potemski R M, Reuter W and Ghez R 1982 *Appl. Phys. Lett.* **41** 1068

Smits F M 1958 *Bell Syst. Tech. J.* **37** 711

Sokolov V I and Shishiyanu F S 1964 *Sov. Phys.—Solid State* **6** 265

Stocker H J 1963 *Phys. Rev.* **130** 2160

Stone L E 1962 *J. Appl. Phys.* **33** 2795

Sturge M D 1959 *Proc. Phys. Soc.* **73** 297

Sze S M and Wei L Y 1961 *Phys. Rev.* **124** 84

Ta L B, Thomas R N, Eldridge G W and Hobgood H M 1983 *Gallium Arsenide and Related Compounds 1982* (Inst. Phys. Conf. Ser. 65) p 31

Tien P K and Miller B I 1979 *Appl. Phys. Lett.* **34** 701

Ting C H and Pearson G L 1971 *J. Electrochem. Soc.* **118** 454

Toyama M 1969 *Jap. J. Appl. Phys.* **8** 1000

Trumbore F A, White H G, Kowalchik M, Luke C L and Nash D L 1965 *J. Electrochem. Soc.* **112** 1208

Tuck B, Adegboyega G A, Jay P R and Cardwell M J 1979 *Gallium Arsenide and Related Compounds 1978* (Inst. Phys. Conf. Ser. 45) p 114

Tuck B 1969 *J. Phys. Chem. Solids* **30** 253

Tuck B 1971 *Phys. Stat. Solidi* (b) **45** K157

Tuck B 1974 *Introduction to Diffusion in Semiconductors* (Stevenage: Peter Peregrinus)

Tuck B 1976 *J. Phys. D: Appl. Phys.* **9** 2061

Tuck B and Adegboyega G A 1979 *J. Phys. D: Appl. Phys.* **12** 1895

Tuck B and Adegboyega G A 1980 *J. Phys. D: Appl. Phys.* **13** 433

Tuck B and Badawi M H 1978 *J. Phys. D: Appl. Phys.* **11** 2541

Tuck B and Badawi M H 1979 *Electronics Lett.* **15** 604

Tuck B and Chaoui R 1983a *Thin Solid Films* **109** 345

Tuck B and Chaoui R 1983b *Electron Lett.* **19** 565

Tuck B and Chaoui R 1984 *J. Phys. D: Appl. Phys.* **17** 379

Tuck B and Hooper A 1975 *J. Phys. D: Appl. Phys.* **8** 1806

Tuck B and Jay P R 1976 *J. Phys. D: Appl. Phys.* **9** 2225

Tuck B and Jay P R 1977a *J. Phys. D: Appl. Phys.* **10** 1315

Tuck B and Jay P R 1977b *J. Phys. D: Appl. Phys.* **10** 2089

Tuck B and Jay P R 1978 *J. Phys. D: Appl. Phys.* **11** 1413

Tuck B and Kadhim M A 1972 *J. Mater. Sci.* **7** 585

Tuck B and Powell R G 1981 *Phys. D: Appl. Phys.* **14** 1317

Tuck B and Zahari M D 1977a *J. Phys. D: Appl. Phys.* **10** 2473

Tuck B and Zahari M D 1977b *Gallium Arsenide and Related Compounds 1976* (Inst. Phys. Conf. Ser. 33a) p 177

Ugandawa T, Higashiura M and Nakanisi T 1980 *Semi-Insulating III–V Materials, Nottingham* ed. G J Rees (Nantwich: Shiva) p 108

Uskov V A 1974 *Sov. Phys.–Semicond.* **8** 1573

Uskov V A and Sorvina V P 1974 *Inorg. Mater.* **10** 895

Van Vechten J A 1982 *J. Appl. Phys.* **53** 7082

Van Vechten J A 1984 *J. Vac. Sci. Technol.* **B2** 569

Vasudev P K, Wilson R G and Evans C A 1980 *Appl. Phys. Lett.* **37** 837

Vieland L J 1961 *J. Phys. Chem. Solids* **21** 318

Vogel R, Dobbener R and Strathmann O 1959 *Z. Metallk.* **50** 130

Watanabe K, Matsuoka Y, Imamura Y and Ito K 1982 *Gallium Arsenide and Related Compounds 1981* (Inst. Phys. Conf. Ser. 63) p 383

Weisberg L R and Blanc J 1963 *Phys. Rev.* **131** 1548

White A M, Porteous P and Dean P J 1976 *J. Electron. Mater.* **5** 91

White A M 1980 *Semi-Insulating III–V Materials, Nottingham* ed. G J Rees (Nantwich: Shiva) p 3

Wieber R J. Gordon J C and Peet C S 1960 *J. Appl. Phys.* **31** 608

Williams E W and Jones C E 1965 *Solid State Commun.* **3** 195

Wilson R G 1982 *Gallium Arsenide and Related Compounds 1981* (Inst. Phys. Conf. Ser. 63) p 1

Wu Y-H, Werner M, Wang S, Flood J and Merz J L 1986 *Electronics Lett.* **22** 115

Yamada M, Tien P K, Martin R J, Nahory R E and Ballman A A 1983 *Appl. Phys. Lett.* **43** 594

Yamamoto Y and Kanbe H 1980 *Jap. J. Appl. Phys.* **19** 121

Yamazaki H, Kawasaki Y, Fujimoto M and Kudo K 1975 *Jap. J. Appl. Phys.* **14** 717

Yee C M, Fedders P A and Wolfe C M 1983 *Appl. Phys. Lett.* **42** 377

Yeh T H 1964 *J. Electrochem. Soc.* **111** 253

Young A B and Pearson G L 1970 *J. Phys. Chem. Solids* **31** 517

Zahari M D 1976 PhD Thesis, University of Nottingham, UK

Zahari M D and Tuck B 1982 *J. Phys. D: Appl. Phys.* **15** 1741

Zahari M D and Tuck B 1983 *J. Phys. D: Appl. Phys.* **16** 635

Zahari M D and Tuck B 1985 *J. Phys. D: Appl. Phys.* **18** 1585

Zotova N V, Lebedev A A and Nasledov D N 1967 *Sov. Phys–Solid State* **8** 1649

Zucca R 1977 *Gallium Arsenide and Related Compounds 1976* (Inst. Phys. Conf. Ser. 33b) p 228

Index

Activity, 33
  coefficient, 37, 40
AlAs
  energy gap, 2
  lattice constant, 3, 6
  melting point, 2
AlGaAs, 4, 211
  /zinc, 107
AlP
  energy gap, 2
  lattice constant, 3, 6
  melting point, 2
AlSb
  /cadmium, 112
  /copper, 190
  energy gap, 2
  lattice constant, 3, 6
  melting point, 2
  /zinc, 105
Antimony
  in GaSb, 204
  in InSb, 203
Arsenic
  in GaAs, 197
  in InAs, 202
Atomic absorption technique, 114
Auger electron spectroscopy, 211

Beryllium in GaP, 114
Boltzmann–Matano technique, 28, 78

Built-in field effect, 38, 190

Cadmium
  in AlSb, 112
  in InAs, 113
  in InP, 108
  in InSb, 114
Capacitance–voltage method, 27
Charged vacancy, 36
Chemical vapour deposition (CVD), 3
Chromium
  in GaAs, 118
  in InP, 156
Cobalt in GaAs, 155
'Constant surface concentration' case, 18
Copper
  in AlSb, 190
  in GaAs, 189
  in InAs, 190
  in InP, 190
  in InSb, 190

Deep level transient spectroscopy, 134
Diffusion
  and mechanical damage, 132
  coefficient, 9
  equations, 9
  -induced dislocations, 82
  -induced disorder, 214, 221

-induced precipitates, 82
ring mechanism, 201
Dip effect, 123, 162, 169, 179
Dislocations, 82, 192, 206
Divacancy, 35, 50, 105
Double-heterojunction laser, 4

Einstein equation, 38
Equilibrium constant, 33
Error function, 15
Error function complement, 17

Fick's first law, 9
Fick's second law, 10
Finite-difference methods, 20
Four-point probe method, 25

GaAs
  /arsenic, 197
  /chromium, 118
  /cobalt, 155
  /copper, 189
  energy gap, 2
  equilibrium vapour pressures, 36
  /gallium, 194
  /gold, 187
  /iron, 153
  lattice constant, 3, 6
  /magnesium, 116
  /manganese, 142
  melting point, 2
  /phosphorus, 199
  /selenium, 61
  /silicon, 69
  /silver, 177
  /sulphur, 47
  /tellurium, 62
  /tin, 63
  /zinc, 78
Ga–As–Cr phase diagram, 122
GaAsP, 200
Ga–As–S phase diagram, 55
Ga–As–Te phase diagram, 55

Ga–As–Zn phase diagram, 87
GaInAs/zinc, 107
Gallium
  in GaAs, 194
  in GaSb, 204
GaP
  /beryllium, 114
  energy gap, 2
  /germanium, 74
  lattice constant, 1, 3, 6
  melting point, 2
  /silver, 187
  /sulphur, 57
  /zinc, 91
Ga–P–Zn phase diagram, 93
GaSb
  /antimony, 204
  energy gap, 2
  /gallium, 204
  /indium, 204
  lattice constants, 3, 6
  melting point, 2
  /tin, 68
  /zinc, 107
Gaussian solution to diffusion
    equation, 12
Germanium in GaP, 74
Gold
  in GaAs, 187
  in InAs, 188
  in InSb, 188
Gunn effect, 1

Hall effect measurements, 25
High electron mobility transistor,
    8

Ideal solution, 34
InAs
  /arsenic, 202
  /cadmium, 113
  /copper, 190
  energy gap, 2

/gold, 188
/indium, 202
lattice constant, 6
melting point, 2
/mercury, 116
/phosphorus, 202
/selenium, 62
/silver, 185
/sulphur, 60
/tellurium, 62
/zinc, 105
Indium
  in GaSb, 204
  in InAs, 202
  in InP, 201
  in InSb, 203
InP
  /cadmium, 108
  /chromium, 156
  /copper, 190
  energy gap, 2
  /indium, 201
  /iron, 156
  lattice constant, 6
  melting point, 2
  /phosphorus, 201
  /selenium, 62
  /silver, 161
  /sulphur, 59
  /zinc, 100, 110
InSb
  /antimony, 203
  /cadmium, 114
  /copper, 190
  energy gap, 2
  /gold, 188
  /indium, 203
  lattice constant, 1, 6
  melting point, 2
  /selenium, 62
  /sulphur, 60
  /tellurium, 62
  /tin, 68

/zinc, 107
Iron
  in GaAs, 153
  in InP, 156
Isoconcentration technique, 31,
    78
  'kick-out' variation, 81

'Kick-out' model, 81

Law of mass action, 32
Light-emitting diodes, 1
Liquid phase epitaxy (LPE), 3
Liquidus, 43

Magnesium in GaAs, 116
Manganese
  in GaAs, 142
  in heavily-doped GaAs, 150
Mercury in InAs, 116
Mole fraction, 34, 42
Molecular beam epitaxy (MBE),
    3, 6
Multiple quantum well, 7, 222

Out-diffusion, 19

Phase diagrams, 42
  Ga–As–Cr, 122
  Ga–As–S, 55
  Ga–As–Te, 55
  Ga–As–Zn, 87
  Ga–P–Zn, 93
Phase rule, 41
Phosphorus
  in InAs, 202
  in InP, 201
  in GaAs, 199
Photoluminescence, 135, 144,
    211, 217
Plasma reflection technique, 51
p–n junction method, 25
Positron lifetime measurements,
    105

Radio-tracer profiling, 24
Ring mechanism, 201

Schottky barrier, 27
Secondary ion mass spectroscopy
    (SIMS), 24
Seebeck coefficient, 27
Selenium
    in GaAs, 61
    in InAs, 62
    in InP, 62
    in InSb, 62
Semi-insulating material, 3, 117
Silicon
    in GaAs, 69
    in superlattices, 216
Silver
    in GaAs, 177
    in GaP, 187
    in InAs, 185
    in InP, 161
Solid solutions, 4
Solidus, 43
    in Ga–P–Zn, 98
Standard state, 33
Stoichiometry, 43
Strained layer superlattices, 210
Substitutional-interstitial
    mechanism, 75
Sulphur
    in GaAs, 47
    in GaP, 57
    in InAs, 60
    in InP, 59
    in InSb, 60
    in superlattices, 217

Superlattice interdiffusion, 210
Superlattices, 210

Tail effect, 204, 206
Tellurium
    in GaAs, 62
    in InAs, 62
    in InSb, 62
Thermo-electric measurements,
    27, 60
Tie line, 43
Tin
    in GaAs, 63
    in GaSb, 68
    in InSb, 68
Type conversion, 144

Vacancy, 35, 53
    charged, 36
    concentration product, 58
    shortfall, 81

X-ray microprobe analysis, 118,
    172

Zinc
    in AlGaAs, 107
    in AlSb, 105
    in GaAs, 78
    in GaInAs, 107
    in GaP, 91
    in GaSb, 107
    in InAs, 105
    in InP, 100, 110
    in superlattices, 214
Zincblende structure, 2

Printed and bound by CPI Group (UK) Ltd, Croydon, CR0 4YY

17/10/2024

01775690-0011